6th International Conference on Microbiome Engineering (ICME 2023)

Berkeley, California, USA
8-10 December 2023

ISBN: 978-1-7138-9155-0

Printed from e-media with permission by:

Curran Associates, Inc.
57 Morehouse Lane
Red Hook, NY 12571

AIChE
120 Wall Street, FL 23
New York, NY 10005-4020

Phone: (800) 242-4363
Fax: (203) 775-5177

www.aiche.org

Additional copies of this publication are available from:

Curran Associates, Inc.
57 Morehouse Lane
Red Hook, NY 12571 USA
Phone: 845-758-0400
Fax: 845-758-2633
Email: curran@proceedings.com
Web: www.proceedings.com

TABLE OF CONTENTS

The Industrialized Gut Microbiome: In Need of Repair

Authors

Sonnenburg, J. - Presenter, Stanford University
The links between our gut microbiome and our health, combined with the malleability of this community, suggests that if we learn the rules for how to manipulate our gut microbes, we may be able to treat and prevent disease. Recent data from our lab and others focused on humans living different lifestyles ranging from hunter-gatherer to industrialized suggest that the gut microbiome is profoundly influenced by lifestyle. Those of us in the industrialized world are harboring a community of microbes that diverges from ancestral states. Loss of species and functions that inhabited humans over much of our evolution, and replacement with new taxa, may result in incompatibilities between the industrialized microbiome and human biology. Diet has emerged as one of the most powerful levers available to shape the composition and function of microbes within the gut. We are addressing whether diet, when combined with bacterial genetic engineering, can be used to engraft desirable microbes into the gut community. Molecular mechanisms of host-microbial interaction are pursued using an array of technologies and experimental approaches including gnotobiotic and conventional mouse models, quantitative imaging, and a metabolomics pipeline focused on defining microbiota-dependent metabolites.

Building the World's Largest Plasmid-Host Interaction Network Using Proximity Ligation Technology

Authors

Auch, B. - Presenter
Liachko, I., Phase Genomics
Sisson, Z., Phase Genomics
Reister, E., Phase Genomics
Mangelson, H., Phase Genomics
Bryson, S., Phase Genomics
Grove, J., Phase Genomics

Plasmids are mediators of horizontal gene transfer (HGT) and serve as both a reservoir and vector for the spread of antimicrobial resistance (AMR) and virulence genes. They are also a valuable tool in synthetic biology, enabling the transformation and engineering of diverse microorganisms. However, plasmid-host relationships in nature have proven challenging to identify without culture-based methods. This culturability requirement restricts the microbial diversity that can be surveyed, and the applicability of these plasmids to engineering of more complex communities.

Proximity ligation sequencing is a powerful metagenomic approach for associating plasmids with their hosts directly in native microbial communities. It captures, *in vivo*, physical interactions between the host microbial genome and the genetic material of plasmids. These linkages offer direct evidence that plasmid sequences were present within an intact host cell; however, unlike culturing experiments, proximity-ligation does not require the propagation of living bacterial cells. The application of this technology to hundreds of complex microbiome samples has yielded thousands of novel plasmid genomes with host assignments, along with large numbers of new microbial genomes. Through broad-scale use of proximity ligation sequencing, we are creating a global-scale database of highly diverse plasmid-host interactions from samples across the world. We will present published and unpublished work highlighting the power of this approach in the field of metagenomic discovery, and its potential in fueling efforts to develop plasmid vectors for microbial engineering across the tree of life.

Spatiotemporal Dynamics during Niche Remodeling By Super-Colonizing Bacteria in the Mammalian Gut

Authors

Urtecho, G. - Presenter, UCLA
Wang, H., Columbia University
Moody, T., Columbia University
Richardson, M., Columbia University
Sheth, P., Sunoco Refining and Marketing
Huang, Y., Columbia University
Ricaurte, D., Columbia University
Velez-Cortes, F., Columbia University
Gerber, G., Brigham and Women's Hospital
Kaufman, A., Columbia University

Fecal Microbiota Transplantation (FMT) has gained traction as a therapy for restoring healthy gut ecosystems within dysbiotic individuals yet has provided mixed results, with outcomes affected by interpersonal variation in microbiome composition. While recent high-profile studies have discovered preliminary markers for successful therapy the specific mechanisms determining the success of microbial transplantation via FMT remain undefined. **Here, we used a mouse gut colonization model that leverages natural inter-individual variation in gut microbiomes to explore the spatiotemporal dynamics of colonizing microbes following FMT.** We show that microbiomes of C57BL/6 mice acquired from different vendors exhibit variable outcomes when subject to inter-vendor FMT and identified a â€œsuper-colonizingâ€ microbiota consortium capable of robustly engrafting in diverse recipients. We then characterized the dynamics of super-colonizers in FMT recipients through a series of longitudinal and spatial metagenomic assays **to identify predictors and mechanisms of successful FMT**. Finally, we used Metagenomic Plot Sampling by Sequencing (MaPS-Seq) to study the micron-scale spatial arrangement of microbes in intact biological samples. By performing MaPS-Seq on FMT communities, we discovered that microbial species form spatially associating communities in the gut that retain their structure after FMT. Furthermore, we found that super-colonization is facilitated by cooperative metabolic activities to exploit unused nutrient niches within the recipient gut. Overall, this study revolutionizes our understanding of microbial dynamics during FMT, revealing previously unknown mechanisms underlying FMT success. These findings have the potential to inform and guide the selection of both donors and recipients, unlocking new approaches to personalized therapeutics.

A bacterium that cannot grow on glucose

Authors

Rakoff-Nahoum, S. - Presenter, Harvard Medical School
The Rakoff-Nahoum lab uses tools of a broad toolkit in microbiology, immunology, host and bacterial genetics, gnotobiotics, bioinformatics and computational biology, centered in ecological and evolutionary frameworks, to understand how microbes interact with the environment, each other and their host. The driving principle of the lab is that by understanding the molecular, genetic, ecological, and evolutionary mechanisms of how microbiome communities form and shape, with focus on the pediatric population, and how the microbiome influences host tissue and physiological systems over time and across health and disease, we can design targeted, ecologically- and mechanistically- rationale interventions to engineer a healthy and normative microbiome-host axis.

Chemistry in the Rhizosphere: Using Exometabolomics and Fabricated Ecosystems to Enable Engineering Microbiomes for Sustainable Agriculture and Soil Carbon Storage

Authors

Northen, T. - Presenter, Lawrence Berkeley National Laboratory
Soil microbiomes are important drivers of plant health and soil carbon cycling. Plants release up to 20% of carbon they fix into the soil surrounding roots (the rhizosphere) to support soil microbes, many of which promote plant health. Overall, 95% of the carbon released by plant roots is returned to the atmosphere through microbial activities. Small-molecule metabolites mediate critical plant-microbe interactions and microbial respiration and can be directly probed using exometabolomics. However, the large variability and lack of control in natural environments makes it difficult to determine causal mechanisms. To address this need, we have developed methods for making and using fabricated plant ecosystems (EcoFABs) complete with synthetic rhizosphere microbial communities. Using these capabilities, we have found identified metabolites that are preferentially used by rhizosphere bacteria and those we predict will accumulate in soils. These findings and approaches help lay a foundation for matching plant hosts with beneficial microbes and harnessing plant and microbial systems for soil carbon storage.

Synbio in the Soil: Tools, Models, and Applications

Authors

Farny, N. - Presenter, Worcester Polytechnic Institute
Many engineered organisms have been developed in recent years for soil-related applications including biosensing, bioremediation, and pathogen control. However, it is unclear how to deploy these solutions safely, as it is difficult to predict survival and persistence of a genetically engineered microbe (GEM) in a complex soil environment. Further complicating the issue, many well characterized synthetic genetic circuits function well within the lab but fail when introduced to the natural soil environment. In order to enable deployment of synthetic biology solutions in soil, two developments are needed: 1) improved genetic circuit designs optimized for function in the soil environment; and 2) enhanced laboratory models of the soil environment that permit automated, high throughput assessment of GEM survival and function. Herein we report on our progress in these two key development areas. We have developed an automatable and scalable liquid soil extract model for monitoring survival and genetic circuit function of an engineered soil organism *Pseudomonas putida*, and have successfully employed this model to examine the relationship of engineered *P. putida* to dominant soil microbes and the native microbiome. We have employed both targeted design and transcriptomics mining approaches to identify both constitutive and inducible promoters with enhanced performance in *P. putida* under a variety of environmental conditions. Our improved tools are expanding our understanding of the relationship of GEMs to the native soil microbiome and advancing knowledge that will enable future deployment of synthetic biology solutions to global soil challenges.

Programming Electronic Communication in Microbial Consortia for Environmental Sensing

Authors

Li, S. - Presenter
Ajo-Franklin, C. M.
Microbial extracellular electron transfer (EET) has been rewired by synthetic biology to develop bioelectronic sensors that can detect environmental stimuli and generate electrical signals. These sensors have been successfully employed to rapidly detect pollutants in the environment. However, prior works have been restricted to a few EET model organisms, limiting the sensor design spectrum and the deployable environments. We recently indicated that the probiotic bacterium *Lactiplantibacillus plantarum* can use the naphthoquinone DHNA secreted by quinone-producing bacteria to perform EET and generate electrical current proportional to DHNA concentrations. Inspired by these discoveries, we develop new bioelectronic sensing strategies by programming the quinone-mediated electronic communication in microbial consortia. We create synthetic co-cultures capable of EET by inoculating *L. plantarum* with a DHNA-producing bacterium, such as *Escherichia coli* or *Lactococcus lactis.* We then engineer DHNA as the signal intermediate for bioelectronic sensing by controlling DHNA synthesis with genetic circuits. When the analyte is present, *E. coli* or *L. lactis* secrete DHNA as the signal amplifier, which mediates EET in *L. plantarum*, resulting in the generation of electrical currents. Using this strategy, we achieve bioelectronic sensing of antimicrobial peptides, organic compounds, reactive oxygen species, and heavy metals. This system holds intrinsic modularity on the strain and genetic levels and can be adapted for sensing other analytes of interest for electrical outputs. Taken together, we demonstrate the potential of rewiring EET in microbial consortia for bioelectronic sensing and communication. These co-cultured based bioelectronic sensors could play a vital role in environmental, food, and health monitoring.

Engineering Synthetic Anaerobic Consortia Inspired by the Herbivore Rumen for Biomass Breakdown and Conversion

Authors

O'Malley, M. - Presenter, University of California-Santa Barbara
Anaerobic microbes work together in complex communities that decompose and recycle carbon biomass throughout the Earth. Despite their importance, little information exists to parse the role of each microbial member within their dynamic community. To address these knowledge gaps, we pioneered new techniques to isolate anaerobes from biomass-rich environments (e.g. guts and fecal materials of herbivores), characterize their shared metabolism, and build synthetic microbiomes to drive biomass to renewable chemicals. Herbivore fecal samples were challenged by different types of biomass during cultivation to identify important microbial partnerships; 10 billion metagenomic reads spread across 402 enrichment samples tracked biological diversity as the cultures converged to a set of stable microorganisms. Nearly 200,000 carbohydrate active enzymes were identified from the fecal samples, and 724 genomes were assembled for previously uncultured microbes. Surprisingly, consortia dominated by anaerobic fungi generated more than twice the amount of methane compared to prokaryotic consortia, suggesting that fungi accelerate biomass breakdown and methane release in herbivores. Overall, our analysis points to natural compartmentalization between anaerobes as a means to degrade crude biomass, which can be exploited to harness natureâ€™s microbes for sustainable chemical production using synthetic systems. For example, by combining anaerobic fungi with chain elongating anaerobic bacteria in synthetic co-culture, we established a cross-feeding system that enables production of short and medium-chain fatty acids directly from lignocellulose. Using analogous bioprospecting approaches, we are currently selecting for plastic-degrading microbial communities in the herbivore rumen, as well as in the natural oil seeps of the Santa Barbara Channel.

Engineering Water Infrastructure Microbiomes: Past, Present, Future

Authors

Nelson, K. - Presenter, U.C. Berkeley
Microbial communities in water infrastructure are leveraged for beneficial purposes, such as degrading waste, while they are also targeted for destruction, to control the transmission of waterborne pathogens. Strategies to engineer these microbial communities were developed long before molecular techniques were available to characterize them. In this talk, I will discuss the evolving understanding of water microbiomes in the essential infrastructure that we rely on to provide sufficient quantities of safe water and manage our waste. I will share specific examples from our research to ensure the safety of "one water" systems that are designed to recycle wastewater directly back into drinking water. We have demonstrated that advanced treatment trains employing a wide variety of technologies can eliminate the original microbial community of the wastewater. However, advanced purified wastewater is not sterile and a new microbiome emerges. In our current research, we explore whether this microbiome can be modified by seeding, and hypothesize that engineering the water microbiome is preferable to relying on chance to ensure that a healthy microbiome is established in this new source of drinking water. I will also touch on significant challenges to managing the water microbiome given inequitable access to these basic services around the globe. I will conclude by speculating on the future of microbiome engineering in water infrastructure systems that aim to be resilient to a changing climate and more equitable.

Easily Accessible Single-Microbe Sequencing (EASi-seq) to Generate Microbiome Atlas.

Authors

Li, X. - Presenter
Abate, A., University of California San Francisco
Xu, L., UCSF
Demaree, B., University of California San Francisco
Noecker, C., University of Washington
Bisanz, J., University of California San Francisco
Weisgerber, D. W., University of Illinois at Urbana-Champaign
Modavi, C., University of California San Francisco
Turnbaugh, P., UCSF
In the latest decade, single-cell sequencing technology has transformed the biology field. However, for the prokaryotic research field, the development of single-cell sequencing techniques has lagged far behind due to various technical challenges. Traditional methods like 16S amplicon sequencing and metagenomics lack single cell resolution and depend on complexed computational algorithms to deconvolute the data. Existing implementations of single microbe genome sequencing use complex workflows that are both labor intensive and require custom microfluid devices. These systems built and operated by well-trained researchers, precluding adoption beyond microfluidic labs. To overcome this issue, we present EASi-seq (Easily Accessible Single microbe sequencing). The EASi-seq protocol enables the accessible shotgun sequencing of individual microbes by adapting the workflow of a commercial single cell instrument originally designed for eukaryotic cell analysis. We successfully applied EASi-seq to the sequencing of a synthetic microbial consortia, a human microbiome sample, and an environmental community. Using the resulting data, we demonstrated the ability to generate detailed single microbe atlases and greatly improve the quality of concurrent metagenomic sequencing. EASi-seq achieved cost-effective single cell microbe sequencing with throughput greater than tens of thousands of cells without the need of specialized expertise with microfluidics. The ability to capture whole genome datasets from thousands of single microbes provides new opportunities for the microbiology research community.

Ultrasound Imaging of Engineered Bacteria Colonizing Tumors and the Gastrointestinal Tract

Authors

Buss, M. - Presenter, California Institute of Technology
Hurt, R., Caltech
Sawyer, D. P., California Institute of Technology
Swift, M., California Institute of Technology
Dutka, P., Caltech
Jin, Z., Caltech
Abedi, M., California Institute of Technology
Shapiro, M., California Institute of Technology

Genetically engineered bacteria have the potential to outperform traditional diagnostics and therapeutics. However, current tools to non-invasively monitor their activity and spatial distribution in vivo are limited. Optical imaging methods suffer from the poor penetration depth of light in tissue, while nuclear imaging methods rely on radioactive tracers. On the other hand, ultrasound is a widely available imaging modality that does not use ionizing radiation and has tissue penetration depth of several centimeters. Recently, the first genetically encoded ultrasound contrast agents based on gas vesicles were developed to link ultrasound to gene expression in bacteria. However, these first-generation acoustic reporter genes (ARGs) expressed too poorly under in vivo conditions to enable ultrasound imaging of bacteria in therapeutically relevant contexts.

Here we present a new and improved ARG construct that produces high levels of robust gas vesicle expression in the probiotic bacterium *E. coli* Nissle (EcN). This second-generation ARG construct, $bARG_{Ser}$, uses genes derived from *Serratia* sp. ATCC 39006 and was optimized for expression in EcN. We demonstrate that with $bARG_{Ser}$, we can use ultrasound to visualize the spatial distribution of engineered EcN after they home to and colonize tumors upon systemic administration. We also demonstrate that the engineered EcN can be imaged with ultrasound when colonizing the gastrointestinal tract and can be used to non-invasively report on biomarkers of gastrointestinal inflammation. By enabling monitoring of engineered probiotic bacteria inside the body, this technology could greatly improve the development and eventual clinical use of this emerging class of microbial cell-based theranostics.

Mapping Gut Microbiota Interactions That Are Robust to C. Difficile strain Variability and Nutrient Landscapes

Authors

Sulaiman, J. E. - Presenter
Thompson, J., University of Wisconsin-Madison
Qian, Y., University of Wisconsin-Madison
Hromada, S., UW Madison
Vivas, E., UW Madison
Safdar, N., University of Wisconsin-Madison
Venturelli, O., University of Wisconsin-Madison
The deleterious human gut pathogen *Clostridioides difficile* strains display extreme genetic variability and confront a changeable nutrient landscape in the gut. By combining high-throughput experiments to build human gut communities with data-driven computational models, we map inter-species interactions shaping the growth of diverse *C. difficile* strains and toxin production in different nutrient environments. We demonstrate *C. difficile* strain-specific variation in the inferred inter-species interaction network and reveal differences in gene expression between *C. difficile* strains. Besides, toxin production of *C. difficile* strains displays large community-context dependent variability, and they are not correlated with growth-mediated interactions. Although *C. difficile* growth is inhibited by gut bacteria in media that promotes high resource competition, most of them cannot inhibit *C. difficile* in media that mimic post-antibiotic perturbation where there are multiple preferred carbohydrates for *C. difficile* to consume. We identify *C. hiranonis* to robustly inhibit the growth and toxin production of diverse *C. difficile* strains across different nutrient environments. Guided by our computational model, we design an anti-*C. difficile* community containing *C. hiranonis* that protects mice from *C. difficile*. Our study highlights strain and nutrient variation as important factors to consider when designing anti-*C. difficile* consortia.

Microbiome-directed Interventions to Improve Pregnancy and Infant Outcomes in the Global South

Authors

Ridaura, V. - Presenter, Bill & Melinda Gates Foundation
Bill and Melinda Gates Foundation microbiome portfolio includes discovery and translational work to understand the microbiome and develop microbiome-directed therapies to be used during pregnancy, lactation and infancy. Our grants on microbiome-directed intervention complement our investments in maternal and infant/child nutrition, including recent work understanding micronutrients and energy requirements in women of reproductive age. Our latest strategy, targeting maternal microbial interventions represents a shift to focus on our growth and resilience principles, to focus on prevention to help restore the microbiota of mothers to improve nutrient absorption, decrease chronic inflammation and improve maternal-fetal health in our focus geographies. Ultimately, maternal microbiome interventions may help improve infant health. Our investments focus on developing novel products for our target populations, by building robust in-country partnerships and facilitating unique partnerships across sectors. Our team is willing to invest in higher-risk ideas to capture the cultural, geographical and biological characteristics of our target geographies to accelerate microbiome-directed product development in the MNCH space.

Formulation Development of Probiotics for Women's and Neonatal Health Applications in Developing Countries

Authors

Heavey, M. - Presenter
Traverso, G., Brigham and Women's Hospital

Several probiotic bacterial strains have been identified as critical players in healthy neonatal development and in maintaining a healthy vaginal microbiome in women. Specifically, it has been observed that infants with low numbers of *Bifidobacterium infantis* are at higher risk for developing premature birth complications, allergy disorders, and reduced weight gain. Further, it has been documented that women with low numbers of *Lactobacillus crispatus* in the vaginal microbiome are at higher risk for contracting a variety of microbial and viral vaginal infections, and are at risk for reduced fertility. There are many challenges associated with administration of these probiotics to the infant gut or to the vaginal tract as these probiotics suffer from high sensitivity to oxygen and harsh environmental conditions. This makes these probiotics difficult to formulate, transport, and administer to patients while maintaining their viability. Here, we use high-throughput excipient library screening, formulation development, *in vivo* pharmacokinetic characterization in sheep (vaginal delivery) and swine (oral delivery), and microbiome analysis to develop highly stable formulations of these probiotics. We explore how lyophilization, prolonged heat exposure, and moisture affect the viability of these probiotics within our formulations. We further characterize how the various physiochemical properties of our formulations impact the *in vitro* release kinetics and the *in vivo* probiotic pharmacokinetics following oral or vaginal delivery. Overall, this work gleans insights into how formulation development strategies can improve the potential for successful administration of highly sensitive probiotic microbes to support healthy neonatal development and to protect women's vaginal and reproductive health.

In Situ Targeted Mutagenesis of Gut Bacteria

Authors

Fernandez-Rodriguez, J. - Presenter

Microbiome research is revealing a growing number of bacterial genes that impact our health. While CRISPR-based tools have shown success in editing human cell genes, their application to bacterial targets within their natural habitat remains a challenge. Our research introduces a pioneering solution by engineering phage-derived particles for *in situ* base editing of *E. coli* inhabiting the murine gut. The engineered particles deliver a base editor encoded in a non-replicative DNA payload, circumventing payload maintenance and dissemination, which allows for both efficient editing and controlled impact on the gut ecosystem. This novel approach achieves remarkable base editing efficiencies, reaching up to 99.7% of the target bacterial population. We first show the stable modification and inactivation of a Î²-lactamase gene in a model *E. coli* strain within the murine gut, persisting for at least 42 days post-treatment. We then leverage this approach to target a pathogenic *E. coli* strain and introduce a mutation in a gene involved in the production of curli, a bacterial appendage implicated in neurodegenerative diseases. Our strategy unlocks the potential for exploration of bacterial gene function within their native environment and presents an opportunity for the development of next-generation microbiome-targeted therapies.

Microbiome Engineering with Synthetic Biology

Authors

Lu, T. K. - Presenter, MIT
Synthetic biology enables the design of diagnostic and therapeutic bacteria that function in the microbiome. For example, transient molecules in the gastrointestinal tract such as nitric oxide and hydrogen sulfide are key signals and mediators of inflammation. Owing to their highly reactive nature and extremely short lifetime in the body, these molecules are difficult to detect. Recently, we published a a miniaturized device that integrates genetically engineered probiotic biosensors with a custom-designed photodetector and readout chip to track these molecules in the gastrointestinal tract (Inda-Webb et al. Nature 2023). Leveraging the molecular specificity of living sensors, we genetically encoded bacteria to respond to inflammation-associated molecules by producing luminescence. Low-power electronic readout circuits integrated into the device convert the light emitted by the encapsulated bacteria to a wireless signal. We demonstrate in vivo biosensor monitoring in the gastrointestinal tract of small and large animal models and the integration of all components into a sub-1.4 cm^3 form factor that is compatible with ingestion and capable of supporting wireless communication. With this device, diseases such as inflammatory bowel disease could be diagnosed earlier than is currently possible, and disease progression could be more accurately tracked. The wireless detection of short-lived, disease-associated molecules with our device could also support timely communication between patients and caregivers, as well as remote personalized care.

Engineering Tools to Control Individual Microbes and Microbiota without Antibiotic Resistance Genes at a Single Strain Level

Authors

Moon, T. S. - Presenter, Washington University in St Louis
Rottinghaus, A., Washington University In St. Louis
Amrofell, M., Washington University in St. Louis
Vo, S., Washington University in St. Louis

Microbial biocontainment is essential for engineering safe living therapeutics [1, 2]. However, the genetic stability of biocontainment circuits is a challenge. Kill switches are among the most difficult circuits to maintain due to the evolution of escape mutants. We engineered two CRISPR-based, chemical- or temperature-inducible kill switches in the probiotic *Escherichia coli* Nissle and demonstrated mutationally robust biocontainment [3]. In this presentation, we will discuss our machine learning-based microbiota engineering tools that are useful to manipulate microbiota and kill pathogens at a single strain level [4]. Specifically, we will discuss the development and validation of a novel computational program, ssCRISPR, which designs strain-specific CRISPR guide RNAs (gRNAs) that can be utilized to modify complex consortia. As a proof of concept, we applied the program to two novel applications: the isolation of specific microbes from consortia through plasmid transformations and the removal of specific microbes from consortia through liposome-packaged CRISPR antimicrobials. Additionally, we will discuss antibiotic resistance gene-free plasmid systems that prevent antibiotic resistance spread via horizontal gene transfer [5]. This new technology has vast implications in designing strain-specific antimicrobials and combating the growing concern of antibiotic- and bacteriocide-resistant microbes.

1. Curr Opin Biotechnol, 2020. 66, 11-17.
2. Biotechnology Journal, 2020. 15, 1900319.
3. Nature Communications, 2022. 13, 672
4. PNAS. 2023. 120, e2213154120
5. Trends. Biotechnol, 2022, 40, 1405-1414

Tracking Horizontal Gene Transfer in Microbiomes Using RNA Memory

Authors

Silberg, J. - Presenter, Rice University
Kalvapalle, P., Rice University
Staubus, A.
Dysart, M., Wyss Institute
Gambill, L., Rice University
Reyes Grandel, K., Rice University
Lu, L. C.
Stadler, L., Rice University
Chappell, J., Rice University

Horizontal gene transfer (HGT) is a primary means by which microbes acquire new DNA and offers a potentially powerful method for microbiome engineering. Despite the potential of HGT for engineering microbes within communities, its dynamics and drivers in natural environments are not well understood. To harness HGT as a gene delivery mechanism, we require tools for in situ quantification and characterization of gene transfer in complex communities. Here, we introduce RNA-Addressable Memory (RAM), an RNA-based genetic memory system that enables high-throughput and quantitative measurements of HGT in microbial communities. RAM uses an engineered ribozyme to attach an RNA barcode onto specific RNA sequences. We encoded RAM on a plasmid such that after entering a recipient cell, RAM generates a ribozyme that targets a conserved region of 16S rRNA and inserts a barcode. Barcoded rRNAs can be amplified to include species-specific variable regions, sequenced, and analyzed to identify transconjugants. We demonstrated the RAM plasmidâ€™s ability to track HGT in a wastewater microbiome. We showed that 140 out of 279 amplicon sequence variants (ASVs) from the sample participated in HGT with an *E. coli* donor carrying a conjugative plasmid. Further, we detected barcoding in low abundance community members, including those present at < 0.1% abundance. RAM can enable cellular programming and biosurveillance of synthetic DNA across diverse environments, deepen our understanding of microbial processes *in situ*, and expand the synthetic biology toolbox for engineering unculturable organisms.

Targeting Drug-Resistant Enteric Bacteria In Vivo with Siderophore Antimicrobial Peptides from Live Engineered Probiotics

Authors

Mortzfeld, B. - Presenter, University of Massachusetts Medical School
Bhattarai, S., University of Massachusetts Chan Medical School
Bucci, V., University of Massachusetts Medical School

The gastrointestinal (GI) tract is the reservoir for multidrug resistant (MDR) pathogens, specifically carbapenem-resistant (CR) *Klebsiella pneumoniae* and other *Enterobacteriaceae*, which often lead to the spread of antimicrobial resistance genes, severe extraintestinal infections, and lethal outcomes. Selective GI decolonization has been proposed as a new strategy for preventing transmission to other body sites and minimizing spreading to susceptible individuals.

Here, we purified the uncharacterized class IIb microcin I47 (MccI47) and demonstrate potent inhibition of numerous *Enterobacteriaceae,* including multidrug-resistant clinical isolates, *in vitro* at concentrations resembling those of commonly prescribed antibiotics. We then genetically modified the probiotic bacterium *Escherichia coli* Nissle 1917 (*EcN*) to produce MccI47 from a stable multicopy plasmid by using MccI47 toxin production in a counterselection mechanism to engineer one of the native *EcN* plasmids, which renders provisions for inducible expression and plasmid selection unnecessary. We then tested the clinical relevance of the MccI47-producing engineered *EcN* in a murine CR *K. pneumoniae* colonization model and demonstrate significant MccI47-dependent reduction of CR *K. pneumoniae* abundance after seven days of daily oral live biotherapeutic administration without disruption of the resident microbiota.

This study provides the first demonstration of MccI47 as a potent antimicrobial against certain *Enterobacteriaceae*, and its ability to significantly reduce the abundance of CR *K. pneumoniae* in a preclinical animal model, when delivered from an engineered live biotherapeutic product. This study serves as the foundational step towards the use of engineered live biotherapeutic products aimed at the selective removal of MDR pathogens from the GI tract.

Competition and Mutualism in the Dysbiotic Vaginal Microbiome

Authors

Dillard, L. - Presenter
Papin, J. A., University of Virginia
Kolling, G., University of Virginia
Glass, E., University of Virginia
Wever, F., Evvy
Thomas-White, K., Evvy
Bacterial vaginosis (BV) is the most prevalent vaginal condition among reproductive-age women experiencing vaginal complaints. Despite its significant impact on women's health, limited knowledge exists regarding the microbial community structure and metabolic interactions associated with BV. In this study, we analyzed metagenomic data obtained from human vaginal swabs to generate *in silico* predictions of BV-associated bacterial metabolic interactions via genome-scale metabolic network reconstructions (GENREs). Our *in silico* simulations revealed significant competition and mutualism of metabolites (L-histidine, selenocysteine, sphinganine-1-phostphate) implicated in host immune function and tissue inflammation. We grew the most common co-occurring bacteria on the spent media of *Gardnerella* species and performed metabolomics to identify potential mechanisms of metabolic interaction. Notably, we identified some BV associated bacteria significantly produce caffeate, a compound implicated in estrogen receptor binding. These findings underscore novel host-microbiome interactions in the dysbiotic vaginal microbiome.

Design of an Orthogonal and Obligate Commensalism for Organism-Based Biological Containment

Authors

Kunjapur, A. - Presenter
Butler, N., University of Delaware
Jones, M., Cornell University
Forti, A., University of Rochester
Biological containment is a critical safeguard for genetically engineered microbes prior to their environmental release to prevent proliferation in unintended regions. However, few biocontainment strategies can support the longer-term microbial survival that may be desired in a target environment without repeated human intervention. In addition, few strategies have been evaluated for their ability to function in the presence of other microbes. Here, we introduce the concept of an orthogonal and obligate commensalism for the autonomous creation of environments that are permissive for survival of a biocontained microbe. We show that this obligate commensalism is highly effective, with the survival of our commensal organism during co-culture dependent on the presence of our producer strain. We also show that this commensalism is orthogonal to a microbial consortium derived from maize root, with survival of the biocontained strain conditional upon the addition of the producer strain to the consortium. Overall, our study demonstrates a transition from a chemical to a biological dependence for biocontained organisms that could lay the groundwork for biocontained synthetic ecologies.

Beyond Containment: Designing Safe Genetically Engineered Microbes (GEMs) for Release

Authors

Abbott, Z. - Presenter, ZBiotics

Advancements in bioengineering made way for the microbial chassis as a factory and delivery vehicle to address those challenges. The predictable outcome is a world where genetically engineered microbes (GEMs) become prevalent not only in modern medicine but also in household pantries, restaurants, and feedstocks. Currently, there are biocontainment avenues to limit the release of GEMs but with the introduction of GEMs specifically intended for release (e.g. probiotics, animal feed additives, soil additives, environmental remediation, etc.) there is little guidance for industry safety standards. We must be prepared to navigate, recognize, and mitigate many unknown unknowns that will inevitably arise with released GEMs. As innovation brings more intended release GEMs to the forefront, it is important that thought-leaders in the community work together to set cautious and thoughtful regulations around GEM safety. At ZBiotics, we aim to coalesce a proactive and scientifically motivated community that fosters transparency and product safety, and we are establishing some early standards with our GEMs that are already available for human consumption. We put forward conscientious engineering that utilizes genes and delivery mechanisms endogenous to the microbe or from other GRAS species to limit novel sites of homology and eliminate the danger of horizontal gene transfer to pathogenic bacteria. Our initiative highlights a method where safety is ingrained in the design. With our experience and vision for ubiquitous adoption of GEMs in daily life, we are engaging the scientific community in a dialog to further refine and adopt intrinsic biosafety measures.

Tuning Interdomain Conjugation Toward in Situ Population Modification in Yeast

Authors

Stindt, K. - Presenter
McClean, M., University of Wisconsin-Madison
The ability to modify and control natural and engineered microbiomes is essential for biotechnology and biomedicine. Fungi are critical members of most microbiomes, yet technology for modifying the fungal members of a microbiome has lagged far behind that for bacteria. Interdomain conjugation (IDC) is a promising approach, as DNA transfer from bacterial cells to yeast enables *in situ* modification. While such genetic transfers have been known to naturally occur in a wide range of eukaryotes, and are thought to contribute to their evolution, IDC has been understudied as a technique to control fungal or fungal-bacterial consortia. One major obstacle to widespread use of IDC is its limited efficiency. In this work, we utilize interactions between genetically tractable *Escherichia coli* and *Saccharomyces cerevisiae* to control the incidence of IDC. We test the landscape of population interactions between the bacterial donors and yeast recipients to find that bacterial commensalism leads to maximized IDC, both in culture and in mixed colonies. We demonstrate the capacity of cell-to-cell binding via mannoproteins to assist both IDC incidence and bacterial commensalism in culture, and model how these tunable controls can predictably yield a range of IDC outcomes. Further, we demonstrate that these lessons can be utilized to lastingly alter a recipient yeast population, by both â€œrescuingâ€ a poor-growing recipient population and collapsing a stable population via a novel IDC-mediated CRISPR/Cas9 system.

A Bacterial Sensor Taxonomy across Earth Ecosystems for Machine Learning Applications

Authors

Park, H. - Presenter
Dehal, P., Lawrence Berkeley National Laboratory
Arkin, A. P., University of California, Berkeley
Joachimiak, M., LBNL
Jungbluth, S., LBNL
Microbial communities have evolved to colonize all ecosystems of the planet, from the deep sea to the human gut. Microbes survive by sensing, responding, and adapting to immediate environmental cues. This process is driven by signal transduction proteins such as histidine kinases, which use their sensing domains to bind or otherwise detect environmental cues and â€˜transduceâ€™ signals to adjust internal processes. We hypothesized that an ecosystemâ€™s unique stimuli leave a sensor â€˜fingerprint,â€™ able to identify and shed insight on ecosystem conditions. To test this, we collected 20,712 publicly available metagenomes from *Host-associated*, *Environmental*, and *Engineered* ecosystems across the globe. We extracted and clustered the collectionâ€™s nearly 18M unique sensory-domains into 113,712 similar groupings with MMseqs2. We built gradient-boosted decision tree machine learning models and found we can classify the ecosystem type (accuracy: 87%) and predict the levels of different physical parameters (R2 score: 83%) using the sensor cluster abundance as features. Feature importance enables identification of the most predictive sensors to differentiate between ecosystems which can lead to mechanistic interpretations. To demonstrate this, a machine-learning model was trained to predict patient disease state and used to identify domains related to oxygen sensing present in a healthy gut but missing in patients with abnormal conditions. Moreover, since 98.7% of identified sensor domains are uncharacterized, importance ranking can be used to prioritize sensors to determine what ecosystem function they may sense. Further, these new predictive sensors can function as targets for novel sensor engineering with applications in biotechnology, ecosystem maintenance, and medicine.

The Gut Microbiota Regulates Host Metabolism through Circadian Clock Genes Nr1d1 and Nr1d2

Authors

Sharma, S. - Presenter
Ma, J., Carnegie Mellon University
Kuang, Z., Carnegie Mellon University
Alterations in the gut microbiota and circadian rhythms of gene expression are known to underlie metabolic disorders. We explored the interplay between the gut microbiota, circadian rhythms, and metabolism by targeting two homologs of the core circadian clock, *Nr1d1* and *Nr1d2* in the mouse intestine. *Nr1d1* and *Nr1d2* are repressed by the gut microbiota. The absence of the gut microbiota leads to decreased relative amplitude of cyclic patterns of these genes. A knockout of *Nr1d1* and *Nr1d2* in the intestinal epithelial cells (IEC) (*Nr1d1/2*$^{î''IEC}$) showed no difference in body weights, glucose and insulin tolerance, but the expression of core circadian clock and metabolic genes was disrupted. Circadian clock genes like *Nr1d1* and *Nr1d2* are known to act on epigenetic modifications to alter gene expression. We found that circadian rhythms of the epigenetic modifications H3K9ac and H3K27ac were disrupted in *Nr1d1/2*$^{î''IEC}$ mice IECs. The global gene expression pattern for *Nr1d1/2*$^{î''IEC}$ and germ-free mice showed lower overall circadian amplitude than that of the wildtype, indicating disrupted circadian rhythms. To further study the role of *Nr1d1* and *Nr1d2* in metabolism, mice were challenged with a high-fat diet, where *Nr1d1/2*$^{î''IEC}$ mice gained more weight, had heavier fat pads, higher fasting glucose and triglycerides levels, and showed impaired glucose and insulin tolerance. Depletion of the gut microbiota during a high-fat diet challenge prevented the previously observed weight gain difference. These findings reveal connections between the gut microbiota, circadian rhythms, and mammalian metabolism.

Towards understanding and designing interactions among critical community members for persistence, activity and protection

Authors

Arkin, A. P. - Presenter, University of California, Berkeley
Specific interactions among different microbes and microbes and their phage can play critical roles in community assembly, activity and persistence. We are developing measurement technologies and approaches to dissect and design these interactions both to understand their role in natural community function and to design therapeutic interventions for applications in protection against microbial infection. We will show how we are combining high throughput genetic, metabolic profiling and scaling phenotyping to create datasets that allow us to infer interactions, illuminate their mechanistic basis, and exploit these interactions for our purposes using examples that span environmental and health applications.

Precision Editing of Non-Model Bacteria in Native Microbiomes Using Mobile CRISPR-Associated Transposases

Authors

Gelsinger, D. - Presenter, Columbia University
Ronda, C., Columbia University
Perdue, T.
Schwanz, L.
Wang, H., Columbia University
Sternberg, S. H., Caribou Biosciences

Deep sequencing shed light on the vast microbial diversity in nature, changing understanding of the crucial roles that microbiomes play in environments. Knowledge of gene functions comes from manipulating the DNA of species in isolation from their natural communities, yet fewer than 1% of microbes are culturable and even fewer are genetically tractable. Moreover, in nature, microbes live in open, dynamic, complex habitats that are difficult to recapitulate in a laboratory. To address these shortcomings, we developed a novel methodology that enables *in vivo* precision microbiome engineering by leveraging highly mobile vectors to effectively shuttle payloads between diverse microbes, with programmable transposases to catalyze targeted DNA integration, a technology we term **MAGICAST**. We optimized delivery and activity of CRISPR-associated transposases in 4 non-model *Bacteroides* species isolated from humans, towards development of a suite of highly mobile gene editing systems. We next introduced novel payloads with high specificity (>90% on-target), using computationally designed metagenomic spacers, and temporal persistence in *Bacteroides* species within the gut microbiome of live mice. We used this to enable genetic tagging of target species *in vivo*, to facilitate their isolation from the environment and investigation into bile acid metabolism. This work defines a new paradigm for genetic studies of gut microbiome and establishes a pipeline for genetic engineering of bacteria from nature into the laboratory. Future work will focus on expanding this technology into other microbiomes (e.g., clinical, extremophilic) to enable discovery of novel functions from undomesticated bacteria with implications in biotechnological, therapeutic, and environmental fields.

Investigating and Leveraging Type I-F CRISPR-Associated Transposon Regulators to Improve Editing Efficiency and Target Range

Authors

Song, L. - Presenter
Swartz, S. - Presenter
Alker, A.
Arora, J.
Rubin, B., University of California
Doudna, J. A., University of California Berkeley
Cress, B.

Type I-F CRISPR-associated transposons (CASTs) are a newly-discovered tool with unique potential in applying targeted genetic edits to complex microbial communities. CASTs are broadly dispersed across bacteria and capable of integrating large genomic payloads. However, little is known about the host molecular factors which regulate CAST integration and their widespread utilization is limited by low efficiency across diverse, non-model bacteria. To expand the range and applicability of Type I-F CASTs as editing tools, we employed a whole genome mutant screening approach to identify putative regulators of CAST transposition in established model systems in which CASTs are known to integrate. Candidate regulator hits were individually validated, with a particular focus on well-characterized genes involved in known mechanisms. Next, we conducted a bioinformatic survey for the conservation of our candidate regulator hits across broad bacterial phyla. Finally, we leveraged our findings by constructing vectors that incorporate these key regulators to increase host range and editing efficiency. These results will shed light on the molecular mechanisms underlying CAST integration and enable more efficient editing in diverse non-model microorganisms.

Rapid design of bacteriophage cocktails to suppress the burden and virulence carbapenem-resistant Klebsiella pneumoniae in the gut microbiome

Authors

Mimee, M. - Presenter, Massachusetts Institute of Technology
Antibiotic usage can lead to expansion of multi-drug resistant pathobionts within the gut microbiome, which can cause life-threatening infections. Alternatives to conventional antibiotics that selectively remove specific taxa are in dire need. Here, we describe a *Klebsiella* PhageBank that enables the rapid design of antimicrobial bacteriophage cocktails to eliminate multi-drug resistant *Klebsiella pneumoniae*. Using a transposon library in carbapenem-resistant *K. pneumoniae*, we identified host factors required for phage infection to inform the design of phage cocktails that mitigate the emergence of resistance. Leveraging the diversity of the PhageBank and experimental evolution strategies, we formulated combinations of phages that effectively eradicate multi-drug resistant *K. pneumoniae in vitro* and in mouse decolonization models. Further, optimized bacteriophage cocktails not only selectively suppressed the burden of multi-drug resistant *K. pneumoniae* in the mouse microbiome, but also selected for acapsular, phage-resistant mutants with attenuated virulence properties. The informed design of lytic phage cocktails could thus prove an effective means to both remove pathobionts and to select for beneficial genotypes in a complex microbial community.

Fungal highways unlock new applications for genetic circuits in the soil microbiome

Authors

Young, E. - Presenter, Worcester Polytechnic Institute
Study of natural soil microbiomes has revealed that bacterial migration and long-distance chemical signaling is facilitated by filamentous fungal highways. These emergent properties in soil are of particular interest for nutrient cycling, pollution remediation, and underground chemical detection, yet they are only just beginning to be understood. In this talk, I will describe a synthetic biology approach to construct a cross-kingdom consortium that enables bacterial migration and signaling with genetic circuits over several centimeters. We first defined a two-member consortium with a genetically tractable, motile bacterium and a filamentous fungal partner. Then, we designed and tested genetic circuits in soil with different chemical inducers, finding that circuits based on quorum sensing functioned best in dry soil. In testing the function of the system in soil columns, we also observe that the presence of the fungal partner accelerates propagation of the engineered bacteria and the quorum sensing signals over several centimeters. These results show that constructed microbiomes are key to achieving desired sensing and signaling functions in soil. Thus, this study builds the foundation for diverse applications of engineered biology in agriculture, nature, and the built environment.

Conjugation as a tool for genetic engineering

Authors

Brophy, J. A. N. - Presenter, Stanford University
Conjugation, *i.e.* the exchange of genetic information from one cell to another through direct contact, is a predominant mechanism of DNA transfer in the wild. It is responsible for the spread of diverse phenotypes in bacterial communities and for the reprogramming of plant cells into proliferative sugar factories. Here, we describe modifications to two different conjugative systems that enable their use as precise genome engineering tools for both plants and soil bacteria.

Engineering Precise Production of Lipopolysaccharides for Intratumoral Immune Activation By Salmonella Typhimurium

Authors

Howell, L. - Presenter
Forbes, N. S., University of Massachusetts
Immunotherapies are promising treatments for cancer; however, many tumors are non-responsive due to immunosuppression. Bacterial therapies demonstrate promise for cancer treatment due to their tumor localization and immunostimulatory properties. Many therapeutic strains of *Salmonella typhimurium* have been attenuated with deletion of the gene *msbB* which catalyzes acylation of surface lipopolysaccharides (LPS). This attenuation allows for safe intravenous injection of the bacteria; but limits their immunotherapeutic efficacy. In this work, we engineered attenuated *Salmonella* to produce msbB upon induction (Sal-msbB) for an inflammatory response once localized to the tumor microenvironment. Utilizing an array of ribosomal binding sites (RBS) and ssrA degradation tags we tuned our system to produce optimal levels of surface LPS to balance safety with therapeutic efficacy. Our system produces 3.75-times as much TNF-Î± once induced, with uninduced bacteria remaining unchanged from attenuated controls. Induced Sal-msbB increased presentation of the activation markers CD80 and CD86 by both monocytes and dendritic cells *in vitro*, which is key for triggering the adaptive immune system. In mice, uninduced Sal-msbB was injected intravenously at therapeutically relevant doses and did not increase systemic cytokine production or induce sepsis after induction, demonstrating a safe delivery profile. Induction of Sal-msbB reduced tumor growth by two-thirds and increased mouse survival versus attenuated controls. Within the tumor, Sal-msbB increases activation of monocytes and production of the pro-inflammatory cytokine IL-6. Overall, the adjuvant activity of controlled production of bacterial LPS increases efficacy of bacterial anti-cancer therapies and provides a natural niche for *Salmonella* therapies in tumor immunotherapy.

Combining Engineered Microbes with Gut-on-a-Chip Systems to Elucidate Disease-Associated Metabolism

Authors

Woolston, B. - Presenter, Northeastern University
Fernez, M., Northeastern University
Hayes, J.
Koppes, A., Northeastern University
Koppes, R., Northeastern University
Lunger, A., Northeastern University
Carrier, R., Northeastern University
Sharma, A., Northeastern University

Engineered commensal microbes are promising platforms for controlling microbial metabolism in the gut microbiota, both for therapeutic outcomes and for fundamental studies aimed at dissecting the complex mechanisms that link microbial activity and disease. This talk will present our recent work engineering *E. coli* to titrate hydrogen sulfide (H_2S) controllably across the physiological range in a gut microphysiological system (gut-on-chip) supportive of the co-culture of microbes and host cells. H_2S is a gaseous microbial metabolite whose role in gut diseases such as IBD and CRC is debated, largely due to the difficulty in controlling its concentration through chemical dosing, and the use of non-representative model systems in previous work. Our gut-on-chip was designed to maintain H_2S gas tension, enable visualization of co-culture in real-time with confocal microscopy, and facilitate sampling for multi-omics characterization. We found that engineered strains colonized the chip and were metabolically active for two days, during which they produced H_2S across a sixteen-fold range and induced changes in host gene expression and metabolism in an H_2S concentration-dependent manner. Interestingly, microbial production led to more highly controlled levels of H_2S than the use of traditional chemical donors. These results validate a novel platform for studying the mechanisms underlying microbe-host interactions, by enabling experiments that are infeasible with current animal and *in vitro* models.

Enabling Targeted Positive Selection in an Anaerobic Lignin-Degrading Bacterial Community

Authors

Alker, A. - Presenter
Rubin, B., University of California
Lawson, C., Lawrence Berkeley National Laboratory

The conversion of plant biomass to biofuel is a promising avenue to reduce the use of fossil fuels. However, the biofuel industry is stymied by the inefficient breakdown of lignin, a complex and uniquely disorganized polymer from plants. While microbial communities present in livestock ruminants and anaerobic digesters may be better adapted to break down biomass more efficiently, controlling these natural communities to maximize degradation and meet industry goals remains a large barrier to this approach. Advancements in microbiome editing have enabled unprecedented access to site-and-species specific targeting of complex microbial communities through the use of RNA-guided CRISPR Cas transposases. However, our ability to enrich edited cells within a population without affecting the surrounding community is limited. In this study, we developed positive selection strategies for anaerobic lignin-degrading microbiomes by searching for broadly indigestible nutrient sources (e.g. C, N, P, metabolites) with complementary small catabolism gene clusters that would be provided on the editing vector. We screened several glucosides for substrate utilization and toxicity when grown in minimal media as the sole carbon source. We performed community amplicon sequencing and metabolite utilization analyses as a proof of principle for the approach. Taken together, this work aims to address a major limitation in our current microbiome editing capabilities by establishing a more universal method for positive selection in lignin-degrading bacterial communities.

Biocontainment Strategies for In Vivo Applications of Saccharomyces Boulardii

Authors

Hedin, K. A. - Presenter
Sommer, M. O. A., Technical University of Denmark
Vazquez Uribe, R., DTU
Kruse, V., Copenhagen University
The human gastrointestinal tract is a complex and dynamic environment, playing a crucial role in human health. Microorganisms engineered to express a therapeutic activity have emerged as a novel modality to manage numerous diseases. Such advanced microbiome therapeutics (AMTs) must be contained within the treated individual. Hence safe and robust biocontainment strategies are required to prevent the proliferation of microbes outside the treated individual. Here we present the first biocontainment strategy for a probiotic yeast, demonstrating a multi-layered strategy combining an auxotrophic and environmental-sensitive strategy. We knocked out the genes *THI6* and *BTS1,* causing thiamine auxotrophy and increased sensitivity to cold, respectively. The biocontained *Saccharomyces boulardii* showed restricted growth in the absence of thiamine above 1 ng/ml and exhibited a severe growth defect at temperatures below 20°C. The biocontained strain was well tolerated and viable in mice and demonstrated equal efficiency in peptide production as the ancestral non-biocontained strain. In combination, the data support that *thi6*âˆ† and *bts1*âˆ† enable biocontainment of *S. boulardii,* which could be a relevant chassis for future yeast-based AMTs.

Development of an orthogonal CRISPR-Cas transposase system for in situ microbial genome editing

Authors

Beckley, J. - Presenter, NC State

Type I-F CRISPR-associated transposon systems have recently been employed for precise, CRISPR RNA-guided insertion of genetic cargo into a variety of chromosomal targets in diverse bacterial species. Although the characterization of novel type I-F CRISPR-associated transposons often involved multi-plasmid assays, the consolidation of these vectors into a single conjugative DNA-editing all-in-one RNA-guided transposase (DART) plasmid has enabled microbial genome editing in a community setting. We demonstrate here the use of transposon Tn7479, derived from *Psychromonas* sp. RZ5, as a novel DART system (PsyDART). Importantly, previous characterization of Tn7479 revealed orthogonality to the original DART system, VcDART, with respect to transposon end recognition and mobilization. Before use as a DART system, Tn7479 was reconstituted as a two-plasmid system and tested with a variety of naturally occurring CRISPR repeat variants to determine the optimal CRISPR RNA sequence for RNA-guided transposition. Additionally, truncated versions of the right and left transposon ends were tested to determine minimal transposon end lengths that retained efficient transposition activity. The optimal CRISPR repeats and truncated transposon ends were used to construct a conjugative suicide PsyDART vector, and validation testing was performed in *E. coli* BL21(DE3). Using an optimized protocol, conjugative delivery by two *E. coli* strains (WM3036 and MFDpir^{+}) resulted in efficient insertion of the associated transposon, encoding an antibiotic resistance gene, at multiple target sites. PsyDART was designed as a broad host range plasmid and future work will be directed towards editing diverse hosts in synthetic soil microbial communities.

Liquid Soil Models for Measuring Survival and Performance of Engineered Pseudomonas Putida

Authors

Carpio Paucar, G. N. - Presenter

Biosensors, bioremediation, and pathogen control using synthetic organisms are among the solutions for soil contamination that have been developed yet, we lack sufficient understanding of how to make the engineered bacteria survive and perform optimally within the complex soil environment. The soil microbial community (SMC), is an essential part of understanding the survival and persistence of genetically engineered microbes (GEM). However, when GEMs encounter the soil, their population is reduced and surpassed by natural organisms. To test the survival of our engineered organism *Pseudomonas putida* in the lab, we created a Soil Extracted Solubilized Organic Matter (SESOM) or sterile liquid soil extract. Over the course of 28 days, we employed 16S rRNA amplicon sequencing to follow changes in soil communities, and we used flow cytometry to examine the dynamic survival of *P.putida* expressing the tdTomato fluorescent marker in sterile and non-sterile SESOM. The taxonomy study revealed that the most prevalent bacterial classes in community soil are *Alphaproteobacteria, Bacteroidia, Actinobacteria, and Gammaproteobacteria*. The samples over the course of 28 days also revealed similarities between solid soil and non-sterile liquid soil extract (NS-SESOM). *P. Putida* growth curve in SESOM was found to be comparable to that of *P. Putida* growing in LB, although there is a population decline when *P. Putida* grows in NS-SESOM. These findings demonstrate that the results obtained in liquid soil extract mirror the solid soil samples. This system is a scalable and automated laboratory model that can forecast GEM survival and persistence, enabling its implementation in the soil.

Mutualism Reduces the Severity of Gene Disruptions in Predictable Ways across Microbial Communities

Authors

Martinson, J. - Presenter
Chacón, J., University of Minnesota
Harcombe, W., University of Minnesota

Predicting evolution in microbial communities is critical for problems from human health to global nutrient cycling. Understanding how species interactions impact the distribution of fitness effects for a focal population would enhance our ability to predict evolution. Specifically, it would be useful to know if the type of ecological interaction, such as mutualism or competition, changes the average effect of a mutation (i.e., the mean of the distribution of fitness effects). Furthermore, how often does increasing community complexity alter the impact of species interactions on mutant fitness? To address these questions, we created a transposon mutant library in *Salmonella enterica* and measured the fitness of loss of function mutations in 3,550 genes when grown alone versus competitive co-culture or mutualistic co-culture with *Escherichia coli* and *Methylorubrum extorquens*. We found that mutualism reduces the average impact of mutations, while competition had no effect. Additionally, mutant fitness in the 3-species communities can be predicted by averaging the fitness in each 2-species community. Finally, the fitness effects of several knockouts in the mutualistic communities were surprising. We discovered that *S. enterica* is obtaining a different source of carbon and more vitamins and amino acids than we had expected. Our results suggest that species interactions can predictably impact fitness effect distributions, in turn suggesting that evolution may ultimately be predictable in multi-species communities.

Programming Bacteria for Multiplexed DNA Detection

Authors

Cheng, Y. Y. - Presenter, University of Wisconsin-Madison
Chen, Z., University of Wisconsin-Madison
Cao, X., University of Wisconsin-Madison
Ross, T. D., University of Wisconsin-Madison
Falbel, T. G., University of Wisconsin-Madison
Burton, B. M., University of Wisconsin-Madison
Venturelli, O. S., University of Wisconsin-Madison

Naturally competent Bacillus subtilis (B. subtilis) can acquire DNA from the environment and incorporate new genetic information. Here we constructed a synthetic genetic circuit in B. subtilis that allows cells to detect extracellular DNA (eDNA) with specific sequences. The modular design of the circuit allows customized DNA input and gene expression output. We built living DNA sensors for detecting diverse major human pathogens including Salmonella typhimurium, Staphylococcus aureus, and Clostridium difficile, characterized sensors' sensitivity and specificity, and achieved multiplexed detection using orthogonal fluorescent reporters. Moreover, the DNA-sensing bacteria can detect target cells without DNA extraction, which is convenient for in vitro or in situ applications. We also characterized the detection efficiency of plasmid and chromosome derived from the donor cells. Plasmid can transfer to *B. subtilis* highly efficiently via the induction of SOS response. Specifically, plasmid can become multimerized and enhance cell lysis in donor cells, which is critical to the efficient horizontal gene transfer. The DNA transfer technology could lead to new applications in the future such as pathogen detection, disease diagnosis, and DNA processing.

Characterizing the Genetic Determinants of Plasmid-Host Interactions

Authors

Arora, J. - Presenter
Rubin, B., University of California
Plasmids shape microbial community diversity by transferring adaptive traits such as resistance, and metabolic capabilities to the microbial hosts. With an increase in sequencing data, plasmid genomes and host associations have become more accessible, but majorly defense genes involved in host compatibility have been identified. Similarly, gene annotation of accessory part of the genomes is a considerable task. To address these gaps, we generated plasmid-bacteria gene interaction networks to identify novel genes interacting between host and plasmid. Additionally, we utilized the co-occurrence frequency to annotate unclassified genes with similar biological functions. To this end, we examined about 6000 plasmid genomes paired with 2500 bacterial genomes in the NCBI nucleotide database to generate bacteria-plasmid networks at the gene level. By examining the plasmid-microbe interactions, this study will provide a mechanistic understanding of host compatibility and help to identify genes that can be targeted for microbiome editing.

Engineering of the Human Skin Bacterium Cutibacterium Acnes for Medical Applications

Authors

Santos-Moreno, J. - Presenter, University of Lausanne
Nevot, G., University Pompeu Fabra
Güell, M., Wyss Institute for Biologically Inspired Engineering
Toloza, L., University Pompeu Fabra
Knödlseder, N., University Pompeu Fabra
A prominent goal of synthetic biology is the development of living medicines based on engineered cells. The genetic modification of the human microbiome presents fewer risks for the host than the editing of human cells, and bacterial therapeutics have been employed to treat metabolic disorders, to limit cancer progression, or to address microbiome dysbiosis. In contrast to the numerous attempts to program gut bacteria for therapeutics, the engineering of the skin microbiome has remained largely unexplored despite its enormous potential.

Cutibacterium acnes, a bacterium that thrives within the sebaceous follicles, is the most abundant human skin microbe and has demonstrated promising engraftment properties when transplanted from a donor to a recipient individual. Despite being an ideal chassis for the development of living skin biotherapeutics, the engineering of *C. acnes* has long been hampered by the extremely low transformation efficiency and the lack of molecular biology tools available.

Here we reverted this situation and rendered *C. acnes* an engineerable chassis for skin disorders. First, we increased the transformability of *C. acnes* by several orders of magnitude. This paved the way for the generation of an engineering toolbox for *C. acnes* that includes modular plasmids and cloning schemes, a promoter library, a library of fluorescent reporters, and gene expression control tools such as CRISPRi and inducible transcription factors. The use of these tools enables the construction of synthetic gene circuits to control *C. acnes* behaviour in a programmable manner, and the development of strains that produce and secrete molecules with therapeutic potential.

Towards Reconstruction of Individual Gut Microbiomes.

Authors

Johns, N. - Presenter
Fischbach, M., Stanford University
Numerous studies have identified gut microbiome compositions and bacterial taxa associated with various host phenotypes of interest. Mechanistic study of how individual strains contribute to these phenotypes in a realistic context has remained challenging due to a lack of tools to precisely edit intact natural communities. An emerging approach to study host-microbe interactions involves the use of complex synthetic communities like hCom2 that have comparable complexity as natural microbiomes (100+ strains) and allow studies that involve the inclusion of exclusion of specific strains. Here we describe our progress towards reconstructing individual microbiomes as modular synthetic communities. By combining high-throughput enrichment and cultivation of strains and long read sequencing we are able to deeply characterize existing phylogenetic diversity within individual samples and identify efficient strategies for obtaining pure cultures of specific taxa. Future efforts will be geared towards reconstruction of microbiomes with transferrable phenotypes of interest related to human health.

Engineered Lactiplantibacillus Plantarum As a Pulmonary Theranostic

Authors

Brasino, M. - Presenter, University of Colorado, Boulder
Merritt, J., Oregon Health and Science University
Wagnell, E., Oregon Health and Science University
Fischer, J., Oregon Health and Science University
Manalo-Hall, E., Oregon Health and Science University

My team and I are developing a platform biosensor technology, consisting of genetically engineered *lactobacillus plantarum* administered to the lower respiratory tract. Using mouse models, we have recently demonstrated that these engineered bacteria may be specifically delivered to the lungs, where they can secrete urinary signals in response to molecules sensed there, and do not colonize or produce a significant adaptive immune response. We have engineered these bacteria to detect pathogens, specific mammalian cells, or damaged tissue. We have also engineered them for therapeutic purposes, by secreting antibiotic peptides and biofilm-degrading enzymes to fight antibiotic-resistant lung pathogens. We will present recent results from the use of these engineered bacteria to detect model lung tumors in immune competent mice.

Metabolic Remodeling of Microbes By Obligate Intracellular Parasites Alters Species Ratios in Microbial Communities

Authors

Bisesi, A. - Presenter
Harcombe, W., University of Minnesota

Bacteria carry a diversity of extrachromosomal genetic elements, including plasmids and some types of bacteriophage (viruses that infect bacteria). Because these genetic elements are obligate intracellular parasites, often dependent on host metabolic processes for replication and proliferation, infection by both plasmids and phage can change host metabolism, driving infected cells into physiological states distinct from uninfected cells. Yet little is known about how infection by these intracellular parasites impacts microbial community composition and function through changes to cellular metabolism. Here we integrate a genome-scale metabolic modeling approach with an *in vitro* system to investigate the consequences of two intracellular parasites of *Escherichia coli* - the plasmid Fâ€™ and the filamentous phage M13 - on interactions between bacteria in a multispecies community composed of cross-feeding strains of *E. coli*, *Salmonella enterica,* and *Methylorubrum extorquens. In silico* experiments suggest that intracellular parasites drive changes in growth rate and cellular excretion that alter community composition in a predictable way across a range of microbial interaction types. *In vitro*, we find that, when bacterial species are engaged in obligate mutualism, infection by either the Fâ€™ plasmid or by both parasites increases species evenness, as predicted by our models. Our work therefore provides a foundation for predicting and managing microbial community composition and function for both biomedical and bioengineering applications.

Engineering Wild Wastewater Microbes to Degrade PET Plastics

Authors

Yip, A. - Presenter
Yip, A. - Presenter
McArthur, O., University of Waterloo
Aucoin, M. G., University of Waterloo
Ingalls, B., University of Waterloo

Microplastics are an emerging contaminant of concern that pose risks to aquatic ecosystems and human health. Wastewater treatment plants are a major pathway for microplastics to enter natural water bodies and soil, and there are currently no unit operations designed to explicitly handle microplastic separation or degradation in wastewater treatment plants. With a focus on PET plastics, we propose a bold solution to removing microplastics in wastewater: engineering wastewater microbiomes to biodegrade PET plastics. In this project, we sought to engineer bacteria from a wastewater sample to produce FAST-PETase, an engineered PET hydrolase that can fully depolymerize PET plastics within approximately one week at mesophilic conditions. We developed a broad-host-range plasmid carrying the FAST-PETase gene and a fluorescent reporter, and we successfully transferred the plasmid to endemic wastewater bacteria using bacterial conjugation. We screened for isolates producing FAST-PETase by purifying the 6X-His-tagged enzyme from the supernatant of concentrated cultures, and then confirmed presence of the enzyme using SDS-PAGE. For each isolate, we screened for activity of the produced FAST-PETase by incubating the enzyme with post-consumer PET plastics for five days. We then characterized PET degradation gravimetrically and confirmed significant mass loss. After confirming expression of FAST-PETase in functional form, we assessed the in situ PET biodegradation rates that each isolate could achieve. Collectively, our work has broad implications regarding the use of engineered microbes for addressing the global plastic pollution problem, and we speculate that our findings may have potential applications in other settings like plastic upcycling and bioremediation.

Multiplexing RNA Addressable Memory to Characterize Synthetic Circuits in Microbial Communities

Authors

Dysart, M. - Presenter, Wyss Institute
Stadler, L., Rice University
Silberg, J., Rice University
Predicting and validating a synthetic circuitâ€™s functionality within non-model microorganisms remains elusive and understudied in microbiome engineering. Although the development of intricate circuits in laboratory strains has become easier and more impressive over the years, application of these advancements into real communities and practical environments has fallen short. To better understand how synthetic parts function in microbes present within communities, we apply and advance a new molecular biology tool, RNA addressable memory (RAM). This tool records which microbial community members receive an engineered plasmid via conjugation. A trans-splicing reaction cleaves the 16S rRNA and attaches a synthetic barcode onto the freshly made edge of the rRNA, forming a chimeric RNA that contains species-specific information (16S) and evidence of horizontal gene transfer (barcode). The identity of transconjugants can easily be read-out using high-throughput amplicon sequencing. In short, RAM records **who** produced a transcriptional event. However, this technology is not limited to just studying horizontal gene transfer. Here, we extend the RAM technology to demonstrate an additional layer of information storage. We show that RAM can be used to store multiplexed information by changing the barcode, addressing synthetic biology questions of â€œWhat, Where, and How?â€™. Unique barcodes are paired with different synthetic parts and deployed in a microbe to understand and predict how the parts function in microbial communities.

Ligand-Specific Biosensing for Aromatic Compounds and Neurochemicals in Engineered Probiotics

Authors

Xi, C. - Presenter, Washington Univeristy In St Louis
Ma, Y., Washington University in St. Louis
Rottinghaus, A., Washington University In St. Louis
Amrofell, M., Washington University in St. Louis
Moon, T. S., Washington University in St Louis
Microbial biosensors have diverse applications in metabolic engineering and medicine. Specific and accurate quantification of chemical concentrations allows for adaptive regulation of enzymatic pathways and temporally precise expression of diagnostic reporters. Although ideal biosensors should differentiate structurally similar ligands with distinct biological functions, such specific sensors are rarely found in nature and challenging to create.

Using *E. coli* Nissle 1917, a â€˜generally regarded as safeâ€™ microbe, we developed and characterized several biosensor systems that promiscuously recognize aromatic amino acids or neurochemicals. To improve the sensorsâ€™ selectivity and sensitivity, we combined rational protein engineering with directed evolution techniques, applicable to both transcription repressors and activators. The generalizable approach involved searching for optimal evolutionary starting points, identifying critical residues in ligand binding, and screening mutagenesis libraries. Our method also provided insights into the previously uncharacterized structures of transcription regulators and elucidated the corresponding specificity control strategies.

We successfully demonstrated the ligand-specific biosensors for phenylalanine, tyrosine, indole-3-acetic acid, phenylethylamine, tyramine, and tryptamine. Each of these structurally similar compounds serves as a distinct biomarker or regulator that influences intestinal functionality and human health. These results lay the groundwork for developing kinetically adaptive microbes for potential applications, ranging from monitoring food quality and detecting diseases to facilitating therapy administration.

The Strain Engineering Platform at the Joint Genome Institute (JGI)

Authors

Liu, H. - Presenter, LBNL
Robinson, D., Lawrence Berkeley National Laboratory: E O Lawrence Berkeley National Laboratory
Zhao, Z., Lawrence Berkeley National Laboratory: E O Lawrence Berkeley National Laboratory
Marimuthu, A., Lawrence Berkeley National Laboratory: E O Lawrence Berkeley National Laboratory
Blaby, I., Lawrence Berkeley National Laboratory: E O Lawrence Berkeley National Laboratory
To understand the molecular mechanisms of the relationship between individual microbes and their community, and their interactions with the environment, we need time-and cost-efficient technologies for efficient genetic manipulation of diverse microbes. To address this challenge, the JGI strain engineering platform focuses on both classical and novel genetic tools development in diverse microbes. First, we adopted Golden-Gate Assembly approach using standard genetic parts to enable fast plasmid vector engineering. Second, we develop novel conjugation transformation donor strains to bypass host restriction-modification systems for efficient DNA delivery. Third, we continuously expand heterologous expression hosts using the chassis-independent recombinase-assisted genome engineering (CRAGE) technology, to enable easy chromosomal integration of large pathways into diverse hosts. As a DOE-funded user facility, we provide various capabilities to users. One major route is to submit a proposal through our Community Science Program (CSP) Annual Call, and there is no cost to the user and no granting of funds.

Deciphering Competitiveness of Root Colonizers with the Use of Synthetic Communities

Authors

Lamouche, F. - Presenter

Plant roots are home to a vast array of microorganisms, collectively known as microbiota. While many of these microbes have no discernible effect on the host, they play a crucial role in maintaining microbial homeostasis at the root-soil interface.

By studying the genomes of entire microbial communities, rather than just individual strains, we can gain access to microbial genetic elements that inform how microbe-microbe and host-microbe- interactions establish.

This approach allows us to identify genetic factors that contribute to niche supremacy, shedding light on the mechanisms by which symbionts and pathogens arise and how homeostasis can be maintained. To investigate, we introduced well-defined synthetic communities (SynCom) composed of 1000 soil-derived, fully sequenced bacteria into Arabidopsis, Lotus, and barley roots. Through metagenome sequencing, we pinpointed 40 isolates dominating specific niches.

Comparing the genomes and conducting Microbial Genome-Wide Association Studies (mGWAS) on these competitive colonizers unveiled molecular functions linked to colonization behavior. Notably, *Lotus japonicus* engages in root nodule symbiosis with *Mesorhizobium loti*. While previous studies focused on single inoculations to identify nitrogen-fixing symbiosis-related genes, our work examined symbiont and microbiome genes contributing to optimal symbiosis and host-soil microbe associations concurrently.

Our analysis highlighted genetic elements enriched in Lotus-compatible commensal bacteria and unique functions in symbiotic rhizobia compared to nonsymbiotic *Mesorhizobium spp.* Together, our studies unravelled novel functions enriched during root nodule symbiosis in a community setting, providing a solid fundament for genetic studies to underpin their role in root nodule symbiosis in the field conditions.

Deployment of a CRISPR-Associated Transposon System to Engineer an Acid Fermenter Microbial Community for Resource Recovery

Authors

LaTurner, Z. - Presenter
Stadler, L., Rice University
Acid fermenters are an emerging biotechnology that harnesses microbial communities to process organic wastes into chemical precursors that can replace many of the compounds traditionally produced by the fossil fuel industry. Synthetic biologists have recently developed tools that give us unprecedented, bottom-up control of microbial communities and which are potentially deployable in acid fermenters to manipulate the often-wide ranging profile of simple, organic products. Better control of this profile will reduce downstream processing costs and increase the efficiency of any ensuing valorization process paired with an acid fermenter.

Specifically, we use conjugation to deliver a recently isolated, transposon-encoded CRISPR-Cas system (INTEGRATE) to native recipients in an acid fermenter community. This system interrupts the acetate pathway causing metabolites to funnel down alternative metabolic pathways, namely the butyrate pathway. INTEGRATE will cause interruption by insertion of a beta-glucosidase gene (*bglC*), into acetate pathway genes. *BglC* results in a protein that can degrade cellobiose, a rate limiting step in the overall degradation of cellulose. Our approach will reduce acetate production, increase butyrate production, and improve degradation of cellulosic wastes in an acid fermenter community.

We demonstrate this by first isolating and expressing *bglC* from *Thermobifida fusca YX* in *E. coli*. Then we incorporate *bglC* into INTEGRATE and use it to interrupt acetate pathway genes in *E. coli*. We next add the customized INTEGRATE system onto a transmissible vector and demonstrate transfer from *E. coli* and successful expression in *Clostridium acetobutylicum*, a common microorganism found in acid fermenter communities.

Predicting Bacteriophage Susceptibility from Escherichia coli genotype

Authors

Moriniere, L. - Presenter
Kazakov, A. E., Lawrence Berkeley National Lab
Svab, M., University of California Berkeley
Listgarten, J., University of California, Berkeley
Dudley, E. G., The Pennsylvania State University
Mutalik, V. K., Lawrence Berkeley National Lab
Arkin, A. P., University of California, Berkeley
The main limitation to the widespread use of bacteriophages (i.e. bacterial viruses) in health, agriculture and biotechnologies relies on the near impossibility to predict the outcome of a specific host-phage interaction. Indeed, mechanisms underlying this remarkable host specificity have been primarily studied in a handful of individual bacterium-phage systems, and an integrative theoretical framework accurately describing the relative influence of host and phage factors remains to be elucidated. Owing to its incredibly diverse pangenome, *Escherichia coli* displays a wide range of phenotypes and is one of key players of the gut microbiome, an environment where it is exposed to a broad range of indigenous phages known as coliphages. In addition, the unparalleled availability of biological and genomic resources for both *E. coli* strains and coliphages makes it the most appropriate model to attempt a systematic, genomics-informed, deciphering of phage susceptibility. The present project aims at developing a machine-learning-driven experimental workflow, which exploits the natural diversity of collections of *E. coli* and coliphages, high-throughput genetic screens and massive susceptibility assays to create a predictive genotype-to-phenotype model. A scalable methodology allowing massive spot assays of phage susceptibility has been developed and successfully applied to an initial set of 75 strains and 95 coliphages, yielding remarkably robust results for over 7,000 single bacterium-phage systems. Cross-comparison of these susceptibility patterns with a range of known important host and phage genetic factors determined through comparative genomics (e.g. phylogenetic relationships, predicted serotype, defense systems...) gave a first glimpse into their relative importance for successful phage infection.

From Genotypes to Ecosystems: Unraveling Microbial Interactions through Machine Learning to Engineer Stable Synthetic Communities

Authors

Kishore, D. - Presenter
Freiburger, A., Argonne National Laboratory
Marsh, A. L., Iowa State University
Cohen, M. B., Iowa State University
Henry, C. S., Argonne National Laboratory
Pelletier, D. A., Oak Ridge National Laboratory
Cottingham, R., Oak Ridge National Laboratory
Doktycz, M. J., Oak Ridge National Laboratory
Ranjan, P., Oak Ridge National Laboratory

Microbial communities are intricate ecosystems of interacting microorganisms in a shared environment. Discerning each dimension of these interactions is crucial to unraveling the principles governing the range of community behavior and their dynamics. However, the complexity and scale of microbial communities lead to a combinatorial explosion of potential community interactions and pose significant challenges for computational modeling and experimental exploration.

In this project, we propose a machine-learning framework that predicts interactions in a microbial community using a suite of novel features derived from the member genotypes. The microbial interaction classifier is being designed to accurately predict the type and strength of interactions between different community members in a given environment. It uses novel scoring metrics incorporating metabolic cooperation and competition, secondary metabolite inhibition, and functional similarity scores extracted from the genomes and reconstructed metabolic models as features. These metrics are designed to capture the different ways in which microbes can interact with each other. The classifier is being trained on a large dataset of microbial interactions, and it is expected to accurately predict interactions between diverse microbes in new environments. Additionally, we aim to utilize this classifier to screen microbial pairings for stable combinations that can constitute a rationally designed synthetic community. This approach can help circumvent experimental limitations and enable controlled experimentation and systematic exploration of microbial community dynamics.

Our machine learning-based approach will provide a valuable framework for predicting microbial interactions, reducing reliance on extensive experimental efforts and accelerating progress in synthetic community design for various applications.

Thermal and Kinetic Modeling of Microbial Reactions during Underground Hydrogen Storage

Authors

Hafedh, A. - Presenter

Underground Hydrogen Storage (UHS) is a promising form of clean energy storage that enables the sustainable use of Hydrogen-Powered Systems on a seasonally-consistent basis. Similar to Carbon Sequestration, a question that arises is how do underground microbial communities react to an increase in hydrogen concentration, as there is still no general consensus nor quantitative effect of the microbial activity on Hydrogen reaction rate. In this study, two models are developed using PHREEQC to observe the biotic redox reactions that occur underground. Specifically, the catalytic effects of bacterial activity in methanogenesis, acetogenesis and sulfate-reduction reactions are observed. The first model is controlled by kinetics and threshold energy thermodynamic constraints, and the second is controlled by a microbial kinetic model. These models are then applied to recorded cases of Underground Hydrogen Storage in literature in order to calibrate models for correct hydrogen reactivity. Through conducting these modeling studies, we can increase confidence in our ability to assess hydrogen reactivity in future Underground Hydrogen Storage (UHS) projects within aquifers or depleted reservoirs.

Design of Orthogonal and Obligate Commensalism for Organism-Based Biological Containment

Authors

Forti, A. - Presenter, University of Rochester
Jones, M., Cornell University
Butler, N., University of Delaware
Kunjapur, A.
Biological containment is a critical safeguard for genetically engineered microbes prior to their environmental release to prevent proliferation in unintended regions. However, few biocontainment strategies can support the longer-term microbial survival that may be desired in a target environment without repeated human intervention. In addition, few strategies have been evaluated for their ability to function in the presence of other microbes. Here, we introduce the concept of an orthogonal and obligate commensalism for the autonomous creation of environments that are permissive for survival of a biocontained microbe. We show that this obligate commensalism is highly effective, with the survival of our commensal organism during co-culture dependent on the presence of our producer strain. We also show that this commensalism is orthogonal to a microbial consortium derived from maize root, with survival of the biocontained strain conditional upon the addition of the producer strain to the consortium. Overall, our study demonstrates a transition from a chemical to a biological dependence for biocontained organisms that could lay the groundwork for biocontained synthetic ecologies.

An Experimental Pipeline for Developing and Optimizing Genetic Tools in Genetic Intractable Gut Bacteria

Authors

Li, J. - Presenter, Stockholm University
The gut microbiota plays a crucial role in human health and disease, shaping both physiological and pathological outcomes in the host. Despite their significance, understanding the genetic underpinnings of their roles is challenging largely due to the lack of robust genetic tools for manipulating these organisms. We unveil an experimental pipeline tailored to craft genetic toolkits for manipulating gut bacteria traditionally deemed resistant to genetic manipulation. Our pipeline utilizes a conjugation-based method and offers detailed guidance to optimize (1) the genetic elements of the conjugative plasmid; (2) the conjugation procedures; and (3) the criteria for transconjugant selection, streamlining the manipulation process for increased simplicity and efficiency. As proof-of-principle for our approach, we applied it to two prominent and highly associated with human health, yet previously genetically intractable gut bacteria, *Akkermansia muciniphila* and *Prevotella copri*. We successfully engineered these microbes for both targeted (plasmid integration) and untargeted (transposon mutagenesis) genetic frameworks. This advancement enables us to (1) refine genetic tools for non-model bacteria, propelling functional genetic investigations in these organisms; (2) render these bacteria amenable for synthetic biology applications, enabling the creation of circuits, sensors, and switches that pave the way for innovative synthetic biology therapeutic interventions capturing ecological and taxonomic diversity in the gut.

Improvements of SYNB1353, an Engineered Bacteria for the Treatment of Homocystinuria, Lead to Increased in Vitro and In Vivo Degradation of Methionine.

Authors

Lubkowicz, D. - Presenter, Synlogic
Means, J.
Bergeron, C., Synlogic
Bruckbauer, S., Synlogic
Renaud, L., Synlogic
James, M., Synlogic Inc.
Schu, J., Synlogic
Perreault, M., Synlogic

Classic homocystinuria (HCU) is an inherited disorder caused by mutations in the cystathionine beta-synthase gene resulting in excessive accumulation of the methionine (Met) degradation product homocysteine (Hcy) and multiorgan clinical manifestations. Early initiation of methionine-restricted diets in HCU patients significantly lowers the risk of developing complications from HCU. To replicate the benefits of Met restriction, the probiotic *Escherichia coli* Nissle 1917 was engineered to generate SYNB1353, a strain designed to degrade Met in the gut and prevent its absorption and conversion to Hcy. SYNB1353 was engineered to facilitate the intake of gastrointestinal Met via a Met importer (MetP) and to convert Met into 3-methylthiopropylamine (3-MTP) via the pyridoxal-5-phosphate (PLP)-dependent Met decarboxylase (MetDC, Q70D N82H). SYNB1353 was evaluated in healthy volunteers and was well-tolerated. Administration of an oral bolus of Met in these subjects resulted in transient elevations of plasma Met and Hcy, which were blunted by SYNB1353 compared to placebo. In preparation for Phase 2 in HCU patients, manufacturing process development continues to improve strain activity. A 2-fold *in vitro* improvement was evaluated in mice treated with d_4-methionine, in which this new fermentation process resulted in increased urinary 3-MTP biomarker levels and significant decrease of plasma d_4-methionine ($p < 0.05$) compared to the old process. Proteomics of SYNB1353 produced with different fermentation processes revealed the presence of a potential PLP competitor that could impact MetDC activity which can be repressed using fermentation conditions. Additional work is ongoing to optimize methionine degradation and SYNB1353â€™s potential for treatment of HCU.

Microbiome Engineering Using Saccharomyces Boulardii

Authors

Kim, J. - Presenter, Korea University Graduate School
Saccharomyces boulardii has great potential for use as a chassis for microbiome engineering due to its high resistance to environmental stress, well-developed genetic tools, and the ability to secrete recombinant proteins in the intestine. As oral feeding of lysozyme has been reported to change the gut microbiome, we engineered *S. boulardii* to secrete human lysozyme, and investigated the changes in the gut microbiome and fecal metabolites in response to the administration of the engineered probiotic yeast into mice. Administration of *S. boulardii* changed the gut microbiome structure by promoting the growth of clostridia and increasing the diversity of strains. The human lysozyme secreted by *S. boulardii* in the intestine resulted in a unique gut microbiome structure through selective growth of specific microorganisms. Our study identified changes in the gut microbiome by administering wild-type *S. boulardii* in mice to healthy mice based on long-read sequencing and demonstrated that a recombinant protein secreted by engineered *S. boulardii* in the intestine could change the microbiome. Our results provide valuable information for the development of bio-therapeutics using engineered *S. boulardii* that changes the gut microbiome and host physiology.

Bifidobacterium 2'-Fucosyllactose Utilization Can be Facilitated By Coexisting Species

Authors

Lou, Y. C. - Presenter

Human milk oligosaccharides (HMOs) ensure proper infant gut microbiome establishment. Isolate studies have revealed the genetic basis for HMO metabolism, but they exclude the possibility of HMO assimilation via synergistic interactions involving multiple organisms. Here, we investigated microbiome responses to 2'-fucosyllactose (2â€™FL), a prevalent HMO and infant formula additive, by establishing individualized microbiomes using fecal samples from three infants as the inocula. *Bifidobacterium breve*, a prominent member of infant microbiomes, typically cannot metabolize 2â€™FL. Using metagenomic data, we predicted that extracellular fucosidases encoded by co-existing members such as *Ruminococcus gnavus* initiate 2â€™FL breakdown, thus critical for *B. breve*â€™s growth. Using both targeted co-cultures and by supplementation of *R. gnavus* into one microbiome, we show that *R. gnavus* can promote extensive growth of *B. breve* through the release of lactose from 2â€™FL. Overall, microbiome cultivation combined with genome-resolved metagenomics demonstrated that HMO utilization can vary with an individualâ€™s microbiome.

Development of a Growth Kinetic Model for Biocontained Members of a Synthetic Consortia

Authors

Jones, M. - Presenter, Cornell University
Forti, A., University of Rochester
Kunjapur, A.
Butler, N., University of Delaware

Intrinsic biological containment approaches offer many advantages in synthetic consortia, such as the ability to control the growth rates and abundance of individual engineered species as well as the mitigation of the risk of unintended proliferation of engineered species in the environment. To work towards ensuring robust biological containment in preparation for environmental release, we developed a growth model to predict the behavior of an intrinsically biocontained microbial strain that is dependent on an orthogonal nutrient for growth. We first demonstrate the tight growth dependence of the strain on a limiting, required metabolite using non-interactive double-substrate limited growth kinetics. To understand the behavior of the biocontained strain in co-culture, we modeled a pairwise interaction of two bacterial strains, one of which is biocontained, and estimated the ability to tune co-culture composition based on the concentration of the required metabolite. Ultimately, we advance towards a quantitative framework for estimating the growth dynamics of a co-culture that features a newly engineered microbial interaction, an obligate and orthogonal commensalism. This model can currently provide qualitative predictions of the impact of inoculation ratio and substrate concentration on the growth of the orthogonally obligate commensal, which should provide a foundation for studying orthogonal biological dependence and biocontained members of synthetic ecologies.

Colonization Persistence of the Live Bacterial Therapeutics Chassis Escherichiacoli Nissle 1917 Deficient of the Colibactin Peptidase (ClbP)

Authors

Neergaard, F. B. F. - Presenter
Kammler, S., DTU Biosustain
Vazquez Uribe, R., DTU
Sommer, M. O. A., Technical University of Denmark
Li, S., DTU BIosustain
Bäckhed, F., The Wallenberg Laboratory, Department of Molecular and Clinical Medicine, Institute of Medicine, Sahlgrenska Academy, University of Gothenburg, Gothenburg, Sweden
Greiner, T. U., The Wallenberg Laboratory, Department of Molecular and Clinical Medicine, Institute of Medicine, Sahlgrenska Academy, University of Gothenburg, Gothenburg, Sweden
Escherichia coli Nissle 1917 (EcN) has a rich history spanning more than a century of use as a probiotic. In recent years, there has been an increasing application of this strain as a platform for the advancement of engineered live therapeutic agents. However, this renewed interest has prompted a reevaluation of the safety aspects surrounding its utility as a live biotherapeutic chassis due to the production of colibactin from the Polyketide Synthase (pks) gene cluster. A multitude of investigations have demonstrated colibactin's genotoxic attributes, showing its ability to alkylate DNA, leading to double-strand breaks. Despite other reports concluding that the mucus layer of the gut can act as a protective barrier against DNA damage, these findings have raised concerns regarding EcN potential association with the development of colorectal cancer. Here, we explore the impact of inactivating the colibactin peptidase (ClbP), which cleaves pre-colibactin into its active form, colibactin, that is associated with its probiotic properties and colonization. Our results reveal that this inactivation has no significant effect on the strain's *in-vitro* characteristics such as growth or antibiotic sensitivity. Furthermore, we demonstrate that the deletion of Î"*clbP* does not adversely affect its ability to colonize mice and does not affect toxicity. Thus, we conclude that this Î"*clbP* deficient strain can be used as a live biotherapeutic.

Exploiting Synthetic Biology and Phage Delivery Technologies to Tackle Antimicrobial Resistance

Authors

Letrari, S. - Presenter
Cimolato, C., University of Padova
Shalata, M. E. M., University of Padova
Del Vecchio, C., University of Padova
Zatta, V., University of Padova
Bernabé, G., University of Padova
Brun, P., University of Padova
Castagliuolo, I., University of Padua
Bellato, M., University of Pavia
Schenato, L., University of Padova
Antimicrobial resistance (AMR) is a health emergency spreading worldwide. The WHO identified the ESKAPE (*Enterococcus faecium*, *Staphylococcus aureus*, *Klebsiella pneumoniae*, *Acinetobacter baumannii*, *Pseudomonas aeruginosa*, and *Enterobacter spp.,* sometimes extended to ESKAPEE to include *Escherichia coli)* among frightens pathogens to be monitored. The request for developing and evaluating new therapies and treatment against AMR is massive and novel approaches must be explored to overcome the onset of new AMR strains. Synthetic biology, combining biological and engineering knowledge, can play a relevant role in this run. From this perspective, this works to exploit phage engineering and CRISPRi technologies to restore antimicrobial resistance in *A. baumanii, K. pneumoniae, and P. aeruginosa*. To achieve this, two main aspects have been explored: (I) the design and optimization of the delivery system, i.e. the engineered phage particles, and (ii) the actuator machinery, requiring the definition of a toolkit/library counting for genetic parts working on the desired strains, also involving the transformation protocols optimization (for decoupling parts characterization from delivery optimization) in non-conventional bacterial models. To assess the efficacy of the actuator machinery, the CRISPRi system, and electroporation protocols were optimized and performed in *A. baumanii* and *K. pneumoniae* obtaining a stable expression of red fluorescent protein (RFP). A transduction assay was performed to evaluate the ability of the engineered phage M13 particles to express the RFP. In this context, further optimization of the delivery platform is needed. In addition, the tropism of M13 phage particles must be studied and explored to infect the target strains.

Understanding the Impact of Intra-Species Bacterial Competition on Native Probiotic Engraftment in the Gut

Authors

Siguenza, N. - Presenter
Tiu, M., UCSD
Brevi, A., UCSD
Mai, I., UCSD
Gootin, H., UCSD
Zarrinpar, A., University of California, San Diego

BACKGROUND: The microbiology field currently assumes that the presence of a specific species in the gut microbiome excludes the subsequent introduction of the same species. Likewise, two different species will not compete for the same niche. However, these assumptions are based on studies in mice with depleted microbiomes. We hypothesize that these findings do not translate to conventional mice.

METHODS: A murine-isolated *Escherichia coli* strain was engineered to express green fluorescent protein (GFP) and kanamycin resistance. The strain was further engineered to express spectinomycin resistance (SpecR) or bile salt hydrolase (BSH) with chloramphenicol resistance. We administered these strains to conventional C57BL/6 mice via oral gavage, varying the timing and order of delivery. We also assessed the influence of different diets (normal chow, high fat, or atherogenic) on colonization outcomes.

RESULTS: Order of administration of the two strains was not the sole predictor of engraftment; prior presence of one did not inherently preclude the latter from colonizing. Additionally, different diets altered the colonization outcomes. Interestingly, colonization by EcAZ-2^{SpecR} or EcAZ-2^{BSH+} is prevented if a mouse is already colonized with a distantly related *Enterococcus* spp.

CONCLUSIONS: We discovered that (a) adding a transgene does not innately reduce native bacterial fitness, (b) the order of bacterial introduction does not solely determine colonization success, (c) diet impacts bacterial competition, and (d) different species can indeed compete for the same niche. These results highlight the complexity of gut microbiome dynamics and emphasize the need for a more nuanced understanding of these interactions.

Effects of Solid-State Fermentation Using R. Oligosporus on the Phytochemical Composition of Wild-Simulated Ginseng Leaf and Its Biological Properties

Authors

Kim, D. - Presenter, Chonnam National University
Lim, J., Seoul National University
Kim, H., Seoul National University
Kim, G. H., University of California San Francisco
Kim, N., Riverside School of Medicine, University of California
Pal, K., National Institute of Technology, Roukela

Wild-simulated ginseng (WSG) harvested from mountainous terrains without specialized facilities is commonly utilized as a functional food in Asian nations due to its abundance of bioactive compounds like ginsenosides. However, the protracted cultivation period, diminished survival rate, and limited size of WSG curtail its application in the food sector. This investigation assessed the fermentation of WSG leaves, which are annually harvestable, employing *Rhizopus oligosporus* over a span of 10 days. The study observed an augmentation in the phytochemical constituents of fermented WSG leaves, inclusive of squalene, phenols, saponins, and a majority of ginsenosides, except Rb1. Remarkably, L-carnitine was newly formed in the WSG leaves, attaining a concentration of 165.1 Î¼g/g post 10 days of fermentation. Additionally, minor ginsenosides Rg2, Rg3, Rh1, F1, F2, and CK demonstrated a 1.32 to 4.43 fold enhancement compared to their unfermented counterparts. Optimal antioxidant activities, gauged by FRAP and ORAC assays, were realized after three days of fermentation. Moreover, fermented WSG leaves exhibited superior inhibition of nitric oxide (NO) release in lipopolysaccharide (LPS)-induced RAW264.7 cells and lipid accumulation in mouse 3T3-L1 cells compared to non-fermented leaves. Consequently, these findings imply that fermented WSG leaves may hold promising applications as functional ingredients within the food industry.

Antimicrobial Compound Production By Engineered Bacteria Against Opportunistic Pathogens in Atopic Dermatitis Skin

Authors

De Sotto, R. B. - Presenter

Atopic dermatitis (AD) is an inflammatory skin condition that is characterized by chronic itching of the skin associated with *Staphylococcus aureus* overgrowth. Current treatment approaches such as topical steroids, antiseptic washes, and biologics are costly, and have undesirable side effects. In this study, a bacterial host is engineered to produce antimicrobial compounds that are effective against *S. aureus*. Since AD skin is prone to secondary infections caused by opportunistic pathogenic commensals, we also explored the ability of these antimicrobial compounds to inhibit opportunistic pathogens, including *Pseudomonas aeruginosa* and *Cutibacterium acnes*. In the presence of the substrate glycerol, our engineered host enabled the production of reuterin, a potent antimicrobial metabolite in the solution. We observed a significant inhibition in all three opportunistic pathogens in the presence of the engineered bacteria as compared to the wild type. Further optimizations will be carried out to ensure the efficacy and safety of these engineered organisms. Ultimately, the effects of the antimicrobial metabolites on the skinâ€™s microbiome will be investigatedâ€”this would uncover whether these metabolites will worsen dysbiosis or restore the abundance of healthy commensals on the skin.

Biochemical Characterization of Novel Hfq-like Proteins in Borg Extrachromosomal Elements

Authors

Cui, M. - Presenter
Alvarado, L., UC Berkeley
Borgs are giant, linear extrachromosomal elements found in anaerobic methane-oxidizing archaea. They feature pervasive tandem repeat (TR) regions that undergo rapid evolution, encoding new TRs within open reading frames and their subsequent proteins. Here, we report the identification of a monophyletic group of novel *Hfq*-like proteins in Borgs genomes. *Hfq* is a pleiotropic regulator of bacterial gene expression, aiding small RNAs to bind with their mRNA targets. Based on phylogenetic and structural bioinformatics, Borg Hfq appears to form hexameric complexes. Strikingly, Borg *Hfqs* display amino acid tandem repeats in their disordered region, which is located at the N-terminal domain (NTD) compared to the C-terminal domain in bacterial *Hfq*. We find through biochemical analysis that bacterial and Borg *Hfq*s display similar RNA binding properties. In-vivo growth experiments show successful rescue of *E. coli Hfq* knockout strains by plasmids with Borg *Hfq*s. Furthermore, NTD-deleted Borg Hfq did not show any negative impact on RNA binding or growth rescue, indicating that the NTD is not necessary for its function. Overall, this study provides insight into the diversity and evolution of Hfq-like proteins and highlights their potential applications in genetic engineering and regulatory processes in microorganisms.

Bacteroides lysogenic Phage for Payload Delivery in the Gut

Authors

Fuerte-Stone, J. - Presenter
Mimee, M., Massachusetts Institute of Technology
Changes in the gut microbiota have been implicated in a variety of disease states. In order to fully investigate the role of the microbiome in these diseases, it is critical to systematically probe the role of each individual species of bacteria in a given consortium. Current methods of probing the microbiome are unable to specifically target strains of interest in these diseases. Bacteriophage are a promising avenue to specifically target strain in a native community due to their narrow host range, antibacterial properties, and ability to lyse host microbes. Here, we propose using lysogenic phage, phage integrated into the host genome, to deliver engineered genetic payloads to hosts of interest in an established microbiome. Using *Bacteroides* as a proof-of-concept target, we identified 14 prophage present in *Bacteroides caccae*. We showed that two of these prophage are capable of transducing DNA to a naïve host by using luciferase and antibiotic markers integrated into the prophage genome. As a proof-of-concept payload, we designed an inducible genetic killswitch using an endogenous *Bacteroides* toxin. This system reduced a targeted bacterial population both *in vitro* and *in vivo*. The combination of the prophage with the killswitch could allow for targeted strain removal in an established microbial community in a mouse gut. Such a technique would allow for evaluation of specific bacterial strains on disease.

Plant Microbiome Engineering for Sustainable Agriculture

Authors

Demirer, G. - Presenter, California Institute of Technology
Berruto, C., Caltech

With a rising human population and the impact of climate change, sustainable agriculture is crucial for meeting societal needs. Traditional methods of enhancing crop productivity through chemical fertilizers and pesticides have led to environmental degradation. Therefore, we must adopt sustainable solutions that decrease our dependence on agrochemicals while boosting farmland productivity. One sustainable approach to enhance plant growth is by using plant growth-promoting rhizobacteria (PGPR), which can improve nutrient use efficiency, provide immunity and growth-stimulating hormones. Current agricultural practices have disrupted and neglected this important microbial community, resulting in a decreased rhizobacteria diversity and dysbiosis of beneficial crop microbiomes. Moreover, the complexity of the preexisting soil microbial community and the variability of soil abiotic conditions create challenges for supplementing desirable soil bacteria into agricultural fields. This complex interaction network between the plant host, environment, and soil microorganisms hinders the development of *designer agricultural microbiomes* with robust plant benefits. Hence, a fundamental understanding of the plant-microbiome communication is essential to re-optimize this relationship.

Plant genotype plays a critical role in biasing the composition of the microbiome, primarily through the secretion of carbon-rich exudates. By altering the exudate composition through plant genetic engineering, we can strategically manipulate and target the microbiome, fostering successful colonization by specific microbes. For this, we developed a workflow identifying plant metabolites in root exudates controlling the colonization and enrichment of various specific PGPR. The final aim is to engineer plant root exudates to enrich functional PGPR solubilizing nitrogen and phosphate under nutrient stress for sustainable plant growth.

A Ribozyme-Based Circuit for Gene Expression Tracking in Non-Model Organisms

Authors

Schwartz, S. - Presenter
Silberg, J., Rice University

Genetic circuits are robust tools for gene expression and have revolutionized our understanding of biological memory and signal actuation. Many of these systems have been developed and confined to model organisms whose Open Reading Frames (ORFs) are well quantified. For emerging areas of microbiome research such as the gut, wastewater, and soil, these circuits are impractical in non-model organisms whose transcriptional components are unknown or poorly understood. However, mRNA serves as an ideal actuator for biological memory, as accessing its information requires no knowledge of ORFs. Tools for manipulating mRNA to produce a measurable output are lacking but have potential to alleviate constraints on host organism for *in vivo* monitoring. Here, we present a simple circuit that uses a programmable guide sequence to direct a trans-splicing ribozyme to convert mRNA into a non-destructive and reversible electrochemical readout. To validate our system, we designed a complementation assay using *E. coli* EW11, a strain incapable of metabolizing inorganic sulfur. We programmed the ribozyme to target mRFP mRNA and output CysI, the hemoprotein subunit of sulfide reductase, enabling production of S^{2-} and survival in minimal media with sulfite as its only sulfur source. Furthermore, bioreactor growth assays using chronoamperometry could discern a S^{2-} readout for over 60 hours. Future research will center on targeting chromosomal genes in non-model organisms as well as develop new parallelizable circuits for synthesis of other redox biomolecules. We hope that this tool can be employed as a novel tool for improved study of microbial systems *in situ*.

Engineered Bacteriophages for in Situ Delivery of CRISPR RNA-Guided DNA Transposases

Authors

Roberts, A. - Presenter
Cress, B.
Adler, B. A., UC Berkeley?UCSF Graduate Program in Bioengineering
Doudna, J. A., University of California Berkeley
Banfield, J. F., University of California
Barrangou, R., North Carolina State University
Bacteriophages are promising vectors for the delivery of engineered DNA to bacteria in diverse microbial communities. Here, we repurposed phage lambda to encode and deliver the DNA-editing all-in-one RNA-guided CRISPR-Cas transposase (DART) system for CRISPR RNA-guided DNA transposition in *E. coli* without lysogeny. For phage engineering, we employed homologous recombination with CRISPR-Cas13a counterselection to replace non-essential regions of the lambda genome, particularly those crucial for lysogeny, with the substantial 10-kb DART system. Consequently, these engineered phages had approximately 25% of their original genome replaced by the DART system, and this modification remained stable across successive passages. The selected lambda phages possess amber mutations and thus rely on amber suppressor *E. coli* hosts for lytic activity and propagation. Therefore, these phages cannot lyse or, due to engineering, establish lysogeny within wild-type *E. coli* hosts. When wild-type *E. coli* strains were infected by DART-encoding engineered phages, site-specific CRISPR RNA-guided integration of the transposon occurred, resulting in simultaneous gene knockout and the introduction of the genes contained within the transposon. Integration events were confirmed via PCR, while whole-genome sequencing verified their specificity and the lack of an associated prophage. Our current efforts focus on broadening target sites and diversifying transposon content for various phenotypic outcomes. By incorporating tailored host-range specificity and ensuring phage infection occurs without lysis or lysogeny, we aim for advancements in delivering the DART system for targeted editing in microbial communities. This work contributes to the development of generalizable phage-mediated genome editing strategies, with future work targeting rhizosphere microbial communities.

Exploring the Landscape of Medium Chain Fatty Acid Production of a Synthetic Anaerobic Bacterial Community Using a Bottom-up Systems Biology Approach

Authors

Hayes, M. - Presenter
Venturelli, O., University of Wisconsin-Madison
Thompson, J., University of Wisconsin-Madison

Microbes exist in complex communities in every environment on earth, leveraging their flexible metabolic capabilities to perform a wide variety of chemical transformations. Anaerobic fermentative communities perform functions of interest collectively, but the species-species interactions which govern the flow of carbon and energy within complex communities are often unknown. The use of computational models informed by experimental data is an emerging technique with applications in understanding microbiome function and designing communities to produce compounds of interest. Medium chain fatty acids (MCFA) are high-value chemicals used in many industries, from pharmaceuticals to textiles, but are currently sourced from fossil fuels. Transforming bioenergy waste streams (biofuel stillage waste) into MCFAâ€™s using microbial bioprocessing holds tremendous potential for both reducing dependency on fossil fuels and extracting additional value-added chemicals from existing bioenergy processes. However, the ecological and molecular mechanisms driving MCFA production remain unresolved. By combining high-throughput experiments and data-driven computational modeling, we investigate the MCFA production landscape of a diverse synthetic anaerobic bacterial community. We utilize an iterative design-test-learn approach to systematically explore the relationship between community composition and MCFA production. Our data-driven approach enables the efficient discovery of high hexanoate producing communities in a landscape of >65,000 possibilities. We confirm *Clostridium kluyveri* as a key MCFA producer, and identify interactions shaping community dynamics. Our models reveal pairwise interactions and environmental parameters governing the favorability of MCFA production, including pH. Our systems biology approach maps the production landscape of industrially relevant compounds and elucidates key control knobs for manipulating MCFA production.

Bacterial Membrane Vesicles Derived from Probiotic B. Subtilis Promote Wound Re-Epithelialization through Engineered Delivery of NOD2 Ligands

Authors

Baerg, L. - Presenter

Bacterial membrane vesicles (MVs) are bilipid nanoparticles secreted as a conserved method of intercellular communication. MVs from Gram-positive bacteria carry diverse bacterial products and represent a unique opportunity in the growing field of postbiotics. This native machinery for mediating microbe-host interactions makes MVs a promising tool for intestinal drug delivery in the context of Crohnâ€™s disease (CD). CD is a chronic inflammatory bowel disease characterized by relapsing inflammation of the digestive tract. CD is associated with (1) loss of function mutations in the host gene encoding the nuclear-binding oligomerization domain 2 (NOD2) pattern recognition receptor and (2) a gut microbiome with reduced capabilities to generate NOD2 stimulating muropeptides from peptidoglycan. However, there are currently no therapeutic strategies targeting NOD2 for CD management. Here, we demonstrate for the first time that MVs from *B. subtilis* can deliver NOD2 ligands *in vitro*. Using HCT116 cells, a human colorectal carcinoma cell line, we demonstrate that treatment with isolated MVs induces IL8 secretion in a NOD2-dependent manner. Furthermore, MVs from *B. subtilis* promote re-epithelialization after *in vitro* scratch wounding. This effect was completely blunted by the addition of an inhibitor for RIPK2, a downstream transducer required for NOD2 signaling. Currently, we are applying metabolic engineering of peptidoglycan degradation and recycling pathways to optimize muropeptide packaging in *B. subtilis* MVs. This work suggests that facilitated delivery of muropeptides by probiotic-derived MVs is a promising strategy to enhance NOD2 signalling activity and promote reconstitution of the epithelial barrier during inflammation-induced injury.

Optimal Production Level of Antibiotics in a Competitive Community

Authors

Tei, M. - Presenter
Arkin, A. P., University of California, Berkeley
To be or not to beâ€"should a microbe put up a fight or conserve resources for a better cause? In a competitive environment with limited nutrients, microbes thrive among each other deploying a few strategies. To maximize its share, it can either enhance its growth or suppress opponentsâ€™ growth. It is postulated that maximal competitive fitness is achieved when there is a balance in resource allocation between self-growth and killing. We developed a theoretical framework to evaluate the cost associated with producing the biomolecular weaponry, as well as the growth advantage from freed-up resources by competitors. We studied a specific case of an engineered community composed of *Bacillus subtilis* and recombinant *Escherichia coli* with *vio* genes from *Janthinobacterium lividum*. Compounds produced by *vio* genes are bacteriostatic against Gram-positive bacteria including *B. subtilis*. We performed simple monoculture experiments to parametrize functions describing cost and benefit stemming from *vio* expression. We then evaluated the theory at predicting *E. coli* growth in cocultures, at various *vio* expression levels and seeding sizes.

Designing Soil Microbiome Function Robust to Environmental Perturbations

Authors

Lee, K. - Presenter
Kuehn, S., University of Chicago
Mani, M., Harvard University
Tikhonov, M., Washington University in St. Louis
Liu, S., Northwestern University

The objective of this project is to ***quantitatively understand how the genomic structure enables soil microbiomes to retain or change emergent metabolic processes in the face of environmental fluctuations***. Microbes seldom exist in isolation and live in complex communities, making their genomic, proteomic, and metabolomic composition highly intricate. Concurrently, these microbial communities experience a barrage of environmental perturbations, such as changes in temperature, pH, moisture, and nutrients. Yet, these communities perform metabolic functions that sustain life in the biosphere, such as global carbon and nitrogen cycling. This raises the question of how these complex systems perform ecosystem functions in the face of environmental perturbations. To address this question, we densely sampled topsoils of pH ranging from 4 to 8 (0.2 increments) from Cook Agronomy Farm (Pullman, WA) and perturbed each soil sample from different native pH to pH of 4 to 8 (13 varying pH levels) in the lab. With both long-term natural pH variation and short-term lab-controlled pH perturbation experiments, we mapped community structure to denitrification activity, a crucial nitrogen cycle process. With simplified consumer-resource models, we observed two modes of functional robustness. In moderate pH perturbations, the community adjusted its dominant species' abundance to adapt to new pH levels. However, in the extreme pH perturbation regimes, the metabolic activity of the community was sustained and even increased due to the emergence of rare native species. Our findings shed light on designing microbial communities with functions robust against environmental perturbations, impacting fields like agriculture and climate change.

Investigating the Impact of Hydrogen Sulfide on the Mucosal Barrier through Development of a Hydrogen Sulfide Biosensor and Multiple Particle Tracking

Authors

Fernez, M. - Presenter, Northeastern University
Carrier, R. L., Northeastern University
Woolston, B., Northeastern University

Inflammatory bowels disease (IBD) is a gastrointestinal (GI) condition that afflicts more than 1% of adults. Hydrogen sulfide, a gaseous compound produced by host and microbial cells, has disputed effects on GI health. While some studies suggest therapeutic effects, others have shown cytotoxicity, mucosal barrier disruption, and promotion of an inflammatory cascade. Resolution of this dispute is hampered by the volatility and reactivity of sulfide, which renders traditional assays inaccurate, and incapable of providing a high-resolution view into spatial and temporal dynamics. To overcome this challenge, we have leveraged genetic sulfide regulation from *Rhodobacter capsulatus* to develop a fluorescent sulfide biosensor in *E. coli*. This system relies on enzymatic conversion of sulfide by sulfide:quinone oxidoreductase (Sqr) to polysulfides, which deactivate the transcriptional repressor SqrR. Through promoter engineering, we optimized the system to provide an operational range of 50 μM - 1 mM under aerobic conditions, but the system was unresponsive under anaerobic conditions due to quinone pool incompatibility. To facilitate sensing in the primarily anaerobic gut microenvironments, we isolated a previously uncharacterized Sqr from *Wolinella succinogenes* that enabled coupling of sulfide oxidation to nitrate and fumarate reduction. Concurrently, using *in vitro* Multiple Particle Tracking, we have shown that sulfide increases diffusivity in porcine mucus. Future work will integrate these two lines of experiment in a gut-on -chip system featuring primary intestinal stem cell monolayer growing on a vertical gel wall for cross-sectional analysis of mucus barriers, enhancing our understanding of the role of sulfide in GI disease.

Machine Learning to Identify Population Specific Microbial Differences in Bacterial Vaginosis

Authors

Parker, I. - Presenter
Ming, D., University of Florida
Celeste, C., University of Florida
Ojo, D., University of Florida

While machine learning (ML) has shown great promise in medical diagnostics, a major challenge is that ML models do not always perform equally well among ethnic groups. Here, we investigated the ability of four ML algorithms to diagnose BV. We determined the fairness in prediction of asymptomatic and symptomatic BV using 16S rRNA sequencing data from Asian, Black, Hispanic, and White women.

Prediction was evaluated using balanced accuracy and average precision of four machine learning models (Logistic Regression (LR), Random Forest (RF), Support Vector Machine (SVM), Multi-layer Perceptron (MLP) classifiers) trained on the dataset.

General purpose ML model performances varied based on ethnicity. Feature selection, using t-test, revealed several bacterial species that varied in accurate prediction. These vary when investigating asymptomatic and symptomatic women. These models can be leveraged to elucidate bacterial species that contribute to adverse BV outcomes and inform novel therapeutic targets.

Developing a Massively Parallel Reporter Assay to Profile in Planta transcriptional Activities of Plant-Associated Bacteria

Authors

Honda, T. - Presenter

Plant roots excrete a variety of metabolites into the rhizosphere and create unique habitats for the microbial communities. Recent metagenomic studies have considerably enriched our genomic knowledge of the plant microbiome. However, much less is studied about the rhizobacterial physiology in association with host plants. One approach to understand it is to measure their transcriptional activities, yet the existing methodologies have limitations: conventional methods such as fluorescent reporters and qPCR are limited to measure a handful of promoter activities, while utilizing genome-wide transcriptomics has been challenging due to a small fraction of bacterial transcripts that are hindered by plant RNA. To this objective, synthetic biology offers an alternative approach. A massively parallel reporter assay (MPRA), which employs DNA barcode as a reporter of promoter activity, allows simultaneous measurements of thousands of promoter activities in a simple workflow. In this study, we developed MPRA in a model rhizobacterium *Pseudomonas simiae* WCS417, leveraging CRAGE (chassis-independent recombinase-assisted genome engineering) to construct a library of strains with barcoded promoters. We established a high-throughput measurement of *in planta* promoter activities of *P. simiae* WCS417 without taking care of plant RNA. Furthermore, we identified a set of uniquely regulated promoters in their process of *Arabidopsis* root colonization and verified these regulated genes substantially influence the root-colonization efficiency. The framework developed here is efficient and scalable, thereby providing an increasing opportunity to understand gene regulation and physiology of rhizobacteria in their host environments.

Genome-Wide Characterization of Diverse Bacteriophages Enabled By RNA-Binding Crispri

Authors

Cress, B. - Presenter
Adler, B. A., UC Berkeley?UCSF Graduate Program in Bioengineering
Al-Shimary, M., University of California, Berkeley
Patel, J., University of California, Berkeley
Schoeniger, J., Sandia National Laboratories
Savage, D., UC Berkeley
Doudna, J. A., University of California Berkeley
Armbruster, E., University of California, San Diego
Pogliano, J., University of California, San Diego
Colognori, D., University of California, Berkeley
Charles, E., University of California, Berkeley
Miller, K. V., University of California, Berkeley
Trinidad, M., University of California, Berkeley
Boger, R., University of California, Berkeley
Nomburg, J., University of California, Berkeley
Beurnier, S., University of California, Berkeley
Cui, M.
Barrangou, R., North Carolina State University
Mutalik, V. K., Lawrence Berkeley National Lab
Lahiri, A., University of California, Berkeley
Bacteriophages play a pivotal role in shaping human health and environmental ecosystems through interactions with bacteria in complex microbial communities. Phages also represent one of the largest reservoirs of unexplored gene content in the biosphere. Even for well-studied model phages, rapid and efficient experimental approaches to identify and investigate their essential genes remain elusive. Here we uncover and exploit the conserved vulnerability of the phage transcriptome to facilitate genome-wide protein expression knockdown via programmable RNA-binding protein dRfxCas13d (CRISPRi-ART) across diverse phages and their host. Establishing the first broad-spectrum phage functional genomics platform, we predict over 90 essential genes across four phage genomes, a third of which have no known function. These results highlight hidden infection strategies encoded in the most abundant biological entities on earth and provide a facile platform to study them. We anticipate that CRISPRi-ART will guide phage engineering strategies for phage therapy and microbiome editing applications.

Engineering Microbial Tryptophan Metabolic Pathways for Controllable Activation of AhR

Authors

McClure, S. - Presenter
Mimee, M., Massachusetts Institute of Technology

The gut microbiota plays a critical role in host physiology. Individual microbes can metabolize components from the diet or produced by the host with their unique set of genetic capabilities. A key mediator in the relationship between host and microbes is tryptophan, originating from exogenous dietary sources. Intestinal microorganisms can metabolize tryptophan into a wide range of downstream metabolites including tryptamine, indole-3-aldehyde, kynurenine, and indole-3-acetic acid. These indole derivatives play a crucial role in regulating host physiology through their interactions with the aryl hydrocarbon receptor (AhR). AhR is crucial for maintaining a functional epithelial barrier and resistance to gut inflammation, in part due to activation of pathways including IL-22 and the production of antimicrobial peptides. Current studies find design challenges with physiological relevance when administering AhR ligands in drinking water or whole bacterium with the ability to produce a wider range of AhR ligands. Our approach will address these issues by engineering single strains containing a biosynthetic pathway for a single AhR ligand to investigate the effects of individual ligands in the most physiological relevant spaces. Altering tryptophan metabolite production within tractable gut commensal, *Bacteroides fragilis,* will allow specific and controllable activation of AhR within the gut microbiome to positively affect host physiology.

Targeted Virulence Reduction and Clearance of Stec Using Magicast

Authors

Perdue, T. - Presenter
Ronda, C., Columbia University
Wang, H., Columbia University
Rivera Gelsinger, D., Columbia University
Schwanz, L.
Shiga toxin-producing *E. coli* (STEC) is a leading cause of foodborne illnesses that currently lacks treatment options. We devised a novel therapeutic approach, MAGICaST, for *in situ* genetic manipulation of STEC that reduces virulence and promotes clearance. MAGICaST utilizes a donor strain to deliver a self-transmissible vector carrying a CRISPR transposon system for precise integration of genetic payloads into target species. Using a mouse model infected with human STEC strain ATCC 43895, we assessed whether disrupting shiga toxin genes, *Stx1b* or *Stx2b*, with integration of a non-coding payload improved host outcomes. Mice administered with wildtype or *Stx1b*-integrated STEC died within five days. However, mice who received *Stx2b*-integrated STEC survived and were asymptomatic over 20 days. Consequently, *Stx2b* was chosen as the target for testing MAGICaSTâ€™s ability to inactivate virulence *in situ*. Three days after administration of the MAGICaST donor, the vector infiltrated 90% of the pathogenic population and inactivated 80% of the virulent *Stx2b* gene, although survival remained unchanged. To promote survival, we used MAGICaST to deliver a genetic payload expressing a secreted TD4 nanobody that inhibits epithelial adhesion by binding the translocated intimin receptor. With integration of the TD4 nanobody into *Stx2b*, survival doubled to ten days. Our results show that MAGICaST is an effective targeted approach to attenuate virulence genes and substitute them with functional payloads to enhance pathogen clearance *in situ*.

Validating a Liquid Soil Model to Explore Soil Microbial Community Dynamics

Authors

Li, S. - Presenter
Farny, N., Worcester Polytechnic Institute
Carpio Paucar, G. N.
Soil microbial communities (SMCs) play an important role in multiple ecological processes, such as plant growth and carbon cycling. However, modeling SMCs in the lab has remained a challenge, because of vast number of unculturable species in soil. Performing assays in solid soil is laborious and time-consuming. Further, solid soil sampling is not amenable to adaptation to high-throughput sequencing platforms. We developed a liquid extraction model, Solubilized Extract of Soil Organic Materia (SESOM), for direct soil metagenomic analysis. We collected DNA from solid soil and liquid-extracted samples and applied 16S rRNA amplicon sequencing to observe the variations of microbe species and abundance in soil communities until after 28 daysâ€™ extraction. Based on the bacteria taxonomy analysis, *Alphaproteobacteria, Bacteroidia*, *Actinobacteria* and *Gammaproteobacteria* are the dominant bacterial classes in all the samples. In addition, the solid SMCs composition was maintained in liquid extraction samples over a period of up to 28 days. SMCs composition in liquid samples represents the starting microbiome in solid samples. The dynamics of solid and liquid samples is further observed by principal coordinate analysis (PCoA) with weighted UniFrac distance analysis of SMCs. A clear separation of two different types of soil in the PCoA plot and different clusters reflect the change of SMC over time. The results promisingly implied that liquid soil extracts could be a good replacement of solid samples to explore and exploit the soil microbial communities for soil-related studies as a more convenient, efficient laboratory model.

Genetically Encoded, Noise-Tolerant, Auxin Biosensors in Yeast Facilitate Metabolic Engineering and Directed Evolution

Authors

Chaisupa, P. - Presenter, Virginia Tech

Auxins are important plant growth regulating compounds that are applied in vast quantities to crops across the globe to control weeds and improve crop quality and yield. Auxins are also produced by nearly every kingdom of life and control both organismal behavior as well as inter-kingdom interactions. Improving our understanding of auxin biosynthesis and signaling is critical to both improving crop plants and controlling symbiotic, commensal, and parasitic inter-kingdom relationships, many of which are critical to ecosystems from forests and oceans to the human microbiome. We present a suite of auxin biosensors that will advance our understanding of and ability to engineer auxin perception by plants and auxin production by fungi. We have developed genetically encoded, ratiometric auxin biosensors in the model yeast *Saccharomyces cerevisiae*, based on the mechanism plants use to perceive auxin. The ratiometric design of these biosensors improves measurements of auxin concentration by reducing clonal and growth phase variation. These biosensors are capable of measuring exogenous auxin in yeast cultures across five orders of magnitude, likely spanning the physiologically relevant range. We implement these biosensors to measure the production of auxin during different growth conditions and phases for *S. cerevisiae*. Finally, we demonstrate how these biosensors could be used to improve quantitative functional studies and directed evolution of plant auxin perception machinery. These genetically encoded auxin biosensors will enable future studies of auxin biosynthesis, transport, and signaling in a wide range of yeast species, as well as other fungi, and plants.

Discovery of Novel Degradation Pathways in Microbial Communities Using Photoaffinity Labeling (PAL)

Authors

Naseem, S. - Presenter

Lignin is the largest renewable aromatic resource and a promising feedstock for sustainable bioproduction. Currently, most efforts are focused on funneling lignin-derived aromatics to common intermediates such as protocatechuate or catechol. The above approach fails to explore the diverse lignin-derived aromatic degradation pathways.

The aim is to explore diverse lignin degradation pathways in nature and discover novel chemical transformations on lignin-derived aromatics by enzymes in microbial communities. This may help us funnel lignin to valuable bioproducts. To discover new pathways, we will use photoaffinity labeling (PAL).

PAL is a biochemical technique that provides information about the ligand-protein interactions in the cell. A lignin-derived aromatic with PAL can be fed to microbial communities. The PAL will label the enzymes using the compound of interest as the substrate. Subsequent enrichment and proteomics will help us identify enzymes processing the labeled aromatics.

When used at high throughput, the workflow will help us explore rich aromatic degradation chemistry in microbial communities and help us funnel the lignin-derived aromatics to valuable bio-products.

The first steps include synthesizing and testing the probes in *Pseudomonas Putida* KT2440, *Achromobacter* isolate, and *Rhodosporidium toruloides*.

Assessing the Effect of Underwear Fabric and Breathability on the Vaginal Microbiome: A Cross-over Interventional Study

Authors

Arizti Sanz, J. - Presenter
Lieberman, T., Massachusetts Institute of Technology
The vaginal microbiome is an important determinant of vaginal health and disease. Epidemiological studies have identified associations between the vaginal microbiome composition and adverse health outcomes, including increased susceptibility to STIs, UTIs and preterm delivery. With an expanding understanding of the vaginal flora's contribution to homeostasis, there arises a potential for developing straightforward, yet effective, interventions for a myriad of conditions. Simple interventions that are easy to implement â€" such as selecting the right underwear fabric â€" hold great promise.

Despite the prevailing conventional wisdom on the benefits of cotton underwear on the vaginal flora, there has been a notable lack of controlled scientific studies to substantiate this claim. To address this, we conducted a cross-over interventional study to examine the effect of clothing breathability on the vaginal microbiome of 30 reproductive-age cisgender women. Participants engaged in two week-long, controlled interventions, during which they self-collected mid-vaginal swabs for in depth analysis, including pH measurements and next generation sequencing.

The results from this study have the potential to open new avenues in designing easy-to-implement interventions that can mitigate the morbidity associated with vaginal dysbiosis and have important implications for womenâ€™s health.

Recording Microbial Signals in Soil: Developing Genetic Memory Devices for Detection of Specialized Metabolites in Microbiomes

Authors

Hansen, M. L. - Presenter
Jelsbak, L., Technical University of Denmark

Microbes reside in complex microbiomes, where secreted specialized metabolites play a major role in interspecies communication. However, investigating the role of specialized metabolites in microbial signaling is currently limited to methods that are disruptive and affected by detection limits. The aim of our research is to engineer soil-dwelling microbes capable of recording the presence of bioavailable specialized metabolites *in situ* in soil and rhizosphere microbiomes. This approach provides an ultrasensitive, non-disruptive alternative to existing methods, which additionally allow for both spatial and temporal investigations of signals *in situ*. The sensors function by inferring an irreversible genetic switch in the presence of the metabolite in question, which is catalyzed by a tightly controlled integrase.

We have developed a handful of different sensors responding to pico-nanomolar concentrations of ecologically relevant metabolites. Each sensor has been tuned to avoid leaky expression in the absence of inducing metabolite in *E. coli*. Sensor sensitivity and responsiveness were analyzed with flow cytometry to generate response curves. Additionally, we have transferred our sensors to the soil- and rhizosphere compatible bacterium *P. putida* KT2440 and determined their effectiveness with flow cytometry. Finally, we have changed the output to genetic memory and verified the efficiency and orthogonality of the sensors with qPCR.

Whole-cell biosensors hold an immense potential to detect and record signals in natural environments. They are highly tunable genetic devices that serve as a promising alternative method for *in situ* investigations of natural products in complex environments such as soil- and plant microbiomes.

Evaluation of Enhanced Skin Probiotics in Organotypic Skin Models

Authors

Nevot, G. - Presenter, University Pompeu Fabra
Santos-Moreno, J., University of Lausanne
Güell, M., Wyss Institute for Biologically Inspired Engineering
Van der Bogaard, E., Radboud University Medical Center
Jansen, P., Radboud University Medical Center
Toloza, L., University Pompeu Fabra

The skin is the largest organ in the human body and serves as the main barrier separating the body from our surroundings. Simultaneously, the skin surface is home to a myriad of microorganisms that play a key role in health. Although many descriptive studies have mapped the skin microbiota composition, novel approaches are needed to elucidate the casual relations among these microbes and the skin. We engineered *Cutibacterium acnes,* the most abundant skin commensal, for its establishment as a synthetic biology chassis to develop tools towards understanding these relations and enable innovative microbiome-based therapeutics. Our *C. acnes* strain harbouring our modular replicative plasmid was tested for survival, applicability and safety in 3D organotypic human epidermal equivalent (HEE) models. The results demonstrated that our strain was able to replicate on the model surface without causing any major inflammatory response. Furthermore, our strain successfully expressed heterologous fluorescent reporters, enabling its imaging in tissue samples. We additionally studied the *in vitro* response of *C. acnes* to stresses such as temperature shock or oxidative stress. By adapting its natural promoters, we developed biosensors capable of a 2.5-fold induction in response to temperature increase or oxidative drugs. Moving forward, we plan to use HEE models with cytokine cocktails to emulate inflammatory conditions and evaluate the capacity of these biosensors to detect diseases biomarkers. In conclusion, we showed that our engineered *C. acnes* is a versatile platform and, together with the HEE model, we have enabled a design-build-test workflow for the development of enhanced skin probiotics.

Investigating and Repurposing Natural Horizontal Gene Transfer Mechanisms for Efficient Microbiome Editing

Authors

Patel, J. - Presenter, University of California, Berkeley
Cress, B.
Doudna, J. A., University of California Berkeley
Ono, R., University of Tokyo
Cruz Sampedro, J., UC Berkeley
Vu, P., UC Berkeley
Adler, B. A., UC Berkeley?UCSF Graduate Program in Bioengineering
Rubin, B., University of California
Diamond, S., UC Berkeley
Kust, A., UC Berkeley

Complex microbiomes employ diverse strategies to control horizontal gene transfer (HGT) and acquisition of new functions. By studying the molecular mechanisms and dynamics of such gene flow, we can both understand the evolution of complex communities and co-opt these mechanisms for precise microbiome editing. To interrogate the factors that can promote HGT in microbial communities, we first synthesize large families of establishment factors and anti-defense genes encoded in mobile genetic elements (MGEs). Then we perform pooled gain-of-function screens to quantify the impact of individual orthologs on conjugation efficiency. Applying this approach to known and putative gene families, we aim to uncover new mechanisms by which MGEs overcome host-encoded barriers to DNA transformation and promote their persistence and dissemination throughout complex ecosystems. In parallel, we have developed a strategy to capture and domesticate plasmids from complex microbiomes, allowing us to harness diverse previously-unaccessed MGE mobilization and replication systems and repurpose them for microbiome editing. Finally, to track the dynamics of MGE dissemination within communities, we are developing a memory-recorder strategy to barcode and record quantitative associations between MGEs and recipient microbes. Together, these strategies should unlock more efficient microbiome editing across diverse taxa and allow us to understand how the MGE arms race impacts horizontal gene transfer in complex microbial communities.

Novel Genetic Tools to Direct Biosynthesis in Intestinal Bacteria

Authors

Arnold, J. - Presenter, Chemical and Biological Engineering, Northwestern University
Recent correlative studies have linked the metabolism of the microbiota to the onset or prevention of various disease states. Complementary advancements in â€˜Omics approaches and anaerobic culture techniques have significantly de-risked harnessing diverse native microbiota members as chassis for implementing synthetic biology strategies. Of great import, gut symbionts, namely butyrate-producing Lachnospiraceae, are thought to play disease-protective roles and provide a unique access point for interventions that modulate host physiology. However, our explicit understanding of how these microbes affect host-facing protective mechanisms remains elusive. This knowledge gap is due to a continued dearth of genome manipulation technologies to control the metabolic activity of health-associated gut bacteria. Here, we propose a blueprint to enable genetic engineering in currently intractable symbionts of the Lachnospiraceae family. Establishing these genetic tools allows for direct genotype-to-phenotype analyses of important â€œbeneficialâ€□ gut symbionts. Additionally, with new physiological insight of these microbes, we can genetically augment living therapeutics and uncover novel microbial metabolites to treat gut-associated disease. To this end, we co-opted plasmid-based systems to establish genetic engineering in Lachnospiraceae. These tools served as a platform to develop a robust genetic toolkit that enables precise manipulation of gene expression in Lachnospiraceae. Moving forward, we will employ these strategies to regulate the production of potentially beneficial metabolites. The development of these tools will aid precise genetic study of many key symbiotic microorganisms, provide a basis to genetically manipulate other intractable microbes, and allow for tuning microbial function in defined contexts to investigate hypotheses that have remained beyond our reach.

Increasing Gut Retention of Probiotics By Targeting Extracellular Matrix Proteins in IBD

Authors

Hazelton, A. - Presenter
Heavey, M.
Nguyen, J., UNC Chapel Hill

Inflammatory bowel diseases (IBD) are chronic, relapsing inflammatory conditions of the intestinal mucosa within the gastrointestinal (GI) tract, mainly Crohnâ€™s disease and ulcerative colitis. Exact pathogenesis of IBD remains unknown, but development is believed to be related to genetics, dysbiotic host-microbe interactions, immunological dysregulation, and the environment. Most IBD therapeutics are anti-inflammatory agents with alterations to the broad immune system and do not directly aim to modulate the gut microbiota. As a result, probiotics, which are viable microbes that exert positive health effects, have been used as therapeutics in GI disorders as they restore the homeostatic conditions of the gut microbiome. Yeast probiotic *Saccharomyces boulardii* is an example of this and is hallmarked by its physiological properties for survival to the lower GI tract that allow for its success as a probiotic. Additionally, *S. boulardii* possess a repertoire of genetic engineering tools allowing for various phenotypic changes that may confer new functions to the microbe. However, like many probiotics, *S. boulardii* has a short gut residence time due to colonization resistance from other gut resident microbes and the harsh conditions of the upper GI tract. This ultimately limits the efficacy of *S. boulardii* to exert its probiotic potential and could thus benefit from an increase in gut retention time. This study explores the use of a targeting moiety using genetic engineering tools to enhance gut retention of the probiotic, *S. boulardii*, to serve as a therapeutic in IBD treatment.

Leveraging Synthetic Biology and Gut-on-a-Chip Systems to Study Gut Metabolite-Host Interactions: A Case Study in H2s

Authors

Hayes, J. - Presenter
Woolston, B., Northeastern University
Koppes, R., Northeastern University
Lunger, A., Northeastern University
Sharma, A., Northeastern University
Koppes, A., Northeastern University

Microbial metabolism in the human gut is increasingly linked with a range of host diseases. Hydrogen sulfide (H_2S) is a gaseous microbial metabolite whose role in gut diseases is debated, with contradictory results stemming from experimental difficulties associated with accurate dosing and measuring H_2S, and the use of model systems that do not accurately represent the human gut environment. Here, we engineered *E. coli* to titrate H_2S across the physiological range in a gut microphysiological system (chip) supportive of the co-culture of microbes and host cells. The chip was engineered to maintain H_2S gas tension and enabled visualization of co-culture in real-time with confocal microscopy. Engineered strains colonized the chip and were metabolically active for two days, during which they produced H_2S across a sixteen-fold range and induced changes in host gene expression and metabolism in an H_2S concentration-dependent manner. These results validate a platform for studying the mechanisms underlying microbe-host interactions.

These findings motivated the development of microbes capable of sequestering or producing intestinal H_2S from different sulfur sources found in the human intestine. First, we engineered a synthetic pathway that converts H_2S to glutathione. The strain consumed H_2S at a rate of 1 mM H_2S/hr-OD_{600}, approximately 10-fold higher than wild type. Next, we engineered a synthetic pathway to convert glutathione to H_2S which can produce 0.3 mM H_2S/hr-OD_{600} compared to no H_2S production in the wild type. These engineered microbes may be used to modify the intestinal metabolome to drive changes in human disease pathology.

Molecular Mechanism for Nutritional Stress Tolerance in Oleidesulfovibrio Alaskensis G20

Authors

Raya, D. - Presenter
Dhiman, S.
The bacterial response mechanism to nutritional stress involves molecular pathways that enable bacteria to adapt and thrive in a nutrient-limited environment. To decipher the underlying molecular mechanisms governing the transition from exponential phase to stationary phase under diminishing nutrient conditions in sulfate-reducing bacteria (SRB), a combinatorial approach of temporal transcriptomics analysis and *ab-initio* simulation, has been deployed in model strain, *e.g., Oleidesulfovibrio alaskensis* G20 (OA-G20), widely documented for microbially induced corrosion. Approximately, 23.7 % genes were consistently upregulated during both the exponential (3rd day) and stationary (9th day) phase, including genes associated with type IV pili, such as *flp and tadE*, exhibiting high upregulation in both conditions. Transcriptomics analysis revealed differential expression of genes linked with stress response, translation and transcription. These results underscore the pivotal role of pili in facilitating the adaption and stress tolerance under nutrient-deficient conditions across growth phases. Through the *ab-initio* analysis, the electronic structure of Flp pili revealed the electrostatic charge distribution along the pili and low electronic HOMO-LUMO gap, indicating its potential involvement in extracellular electron transfer. The electronic structure further substantiates the significance of C-terminal region of pili in catalyzing the last step of electron transfer. An in-depth analysis of Flp pili biogenesis and electronic properties will aid in engineering future synthetic microbiomes for biomanufacturing applications as well as have larger insights into biocorrosion research.

Biomanufacturing in Space: Bionutrients and CO2-Based Manufacturing

Authors

Paddock, M. - Presenter
Ball, N., KBR
Soundararajan, M., KBR
Kagawa, H., KBR
Hindupur, A., NASA Ames Research Center
Vu, S. T., Nasa Ames Research Center
Kostakis, A., NASA Ames Research Center
Sims, K., NASA Ames Research Center
Sharif, S., NASA Ames Research Center
Villanueva, A. G., NASA Ames Research Center
Roque, O. A., NASA Ames Research Center
Downing, S. A., NASA Ames Research Center
Dougherty, M. J., NASA Ames Research Center
Galazka, J. M., University of California at Berkeley
Ray, H. E., NASA Ames Research Center
Anderson, L. M., ASRC Federal
Jones, H. W., NASA Ames Research Center
Gresser, A. L., NASA Ames Research Center
Settles, A. M., NASA Ames Research Center
Hogan, J. A.
Donovan, F. M., NASA Ames Research Center
Biomanufacturing can provide on-demand production of mission-critical compounds and materials to support long-duration space exploration while circumventing the challenges of transporting materials from Earth. The NASA Synthetic Biology Project is developing two biomanufacturing capabilities: BioNutrients and CO_2-Based Manufacturing. BioNutrients is an ongoing mission aboard the International Space Station focused on the production of perishable nutrients in an on demand for direct for consumption. The first flight experiment of this project targeted the production of carotenoids: Î^{2}-carotene and zeaxanthin, in recombinant yeast strains. Since then, the project has expanded to encompass the production of the fermented consumables like yogurt and kefir for use as a nutrient delivery mechanism. The CO_2-Based Manufacturing system aims to use *in situ* resources to allow for biomanufacturing with minimal re-supply required. The manufacturing platform is combined with an electrochemical CO_2 conversion system which can produce simple carbon substrates to support microbial based biomanufacturing. A comprehensive ground-based platform for recombinant protein purification is in development with the goal of producing a thermostable carbonic anhydrase enzyme from *E. coli* utilizing CO_2-derived acetate. Our group hopes to propel advancements in space biomanufacturing for long duration space flight by harnessing the tools of synthetic biology.

Microbiome of the Termite Gut and Its Biotechnological Perspective

Authors

Calusinska, M. - Presenter
Klimek, D., Luxembourg Institute of Science and Technology
Bertucci, M., Luxembourg Institute of Science and Technology
Goux, X., Luxembourg Institute of Science and Technology
Lignocellulose is the most abundant renewable biopolymer on Earth, and its deconstruction is observed under a wide range of environmental conditions. Diverse organisms have evolved specialised strategies and enzymes, with termites being the most efficient natural lignocellulose degraders.

The â€œhigher termiteâ€ gut is a complex multi-domain system where symbiosis between prokaryotes and the host has established over the evolutionary timescale. The gut microbiome is mainly dominated by Firmicutes, Spirochaetes, Fibrobacteres and Bacteroidetes representatives, most of which are yet uncultured and genetically intractable microorganisms. Our recent metatranscriptomic and metagenomic studies combined with genome re-constructions have showed that only a minor fraction of microbial carbohydrate active enzymes (CAZymes) is shared between the different termite gut microbiomes, making each termite species a unique reservoir of potentially industrially applicable enzymes. As an example, we characterised several GH5_4 multi-functional CAZymes of Spirochaetes origin, which can simultaneously attack cellulose, xylan and mannan backbones. Broad substrate specificity CAZymes are potent and versatile tools for a renewable bioeconomy and could help simplifying the complexity of enzymatic cocktails. In additional to being host-specific, microbial CAZyme gene expression is also gut compartment-specific, showing further niche specialisation of certain microbial populations. For some termite species, the axial and radial gradients of various physical and chemical parameters are very pronounced between the different gut segments, allowing specialisation in efficient utilisation of separate lignocellulose fractions.

Further characterisation and future engineering of the termite gut-compartment specific microbiomes will allow establishing best microbial consortia with enzymatic machinery efficiently processing industrially important lignocellulosic biomasses.

Metabolic Complexity Drives Divergence in Microbial Communities

Authors

Silverstein, M. - Presenter
Segrè, D., Boston University
Bhatnagar, J., Boston University
Microbial communities are shaped by the metabolites available in their environment, but the principles that govern whether different communities will converge or diverge in any given condition remain unknown, posing fundamental questions about the feasibility of microbiome engineering. To this end, we studied the longitudinal assembly dynamics of a set of natural microbial communities grown in laboratory conditions of increasing metabolic complexity. We found that different microbial communities tend to become similar to each other when grown in metabolically simple conditions, but diverge in composition as the metabolic complexity of the environment increases, a phenomenon we refer to as the divergence-complexity effect. A comparative analysis of these communities revealed that this divergence is driven by community diversity and by the diverse assortment of specialist taxa capable of degrading complex metabolites. An ecological consumer-resource model of community dynamics indicates that the hierarchical structure of metabolism itself, where complex molecules are enzymatically degraded into progressively smaller ones, is necessary and sufficient to recapitulate all of our experimental observations. In addition to pointing to a fundamental principle of community assembly, the divergence-complexity effect has important implications for microbiome engineering applications, as it can provide insight into which environments support multiple community states, enabling the search for desired ecosystem functions.

How to Drive the Transmission of Seed Bacterial Communities to Seedling

Authors

Sarniguet, A. - Presenter
Simonin, M., INRAE
Arnault, G., INRAE
Garin, T., INRAE
Marais, C., INRAE
Brin, C., INRAE
Brault, A., INRAE
Briand, M., INRAE
Barret, M., INRAE
Seeds represent the starting point of plant microbiota assembly, which includes commensals, pathogens and beneficial microbes. The seed is therefore a key vector organ for plant health and protection. However, harnessing beneficial plant-microbiota interactions requires exploring and understanding the processes involved in microbiota assembly on seeds and its modifications during germination and plant emergence. A prerequisite was to describe the importance of seed microbiota in a large-scale meta-analysis across 50 plant species that shows a core microbiota composed of 30 microbial taxa and a very diverse flexible microbiota (Simonin et al. 2022). However, our results suggest that the most abundant seed microorganisms are not necessarily the ones that have the highest transmission success to seedlings. Our studies on bean, radish and rapeseed show that plants first recruit rare bacterial taxa from soil and seed (Rochefort et al 2019, Chesneau 2022). To better understand the transmission and effects of seed-borne taxa, we then performed inoculations of seed-borne bacterial synthetic communities (SynComs) on seeds. These experiments demonstrated a phylogenetic signal associated with the transmission success of strains from seeds to seedlings and the important role of mass effects. Additionally, we observed a strong influence of the composition of the initial SynCom on transmission success due to biotic interactions. SynCom inoculations also led to a modulation of various plant phenotypes, including emergence rates, seedling development and impact of a soil-borne pathogenic fungus. Altogether, we present evidence of the feasibility and potential of plant microbiota engineering using seed-borne bacterial SynComs inoculated on seeds.

Understanding Transcriptional Regulatory Units in Pseudomonas Putida Under Changing Environmental Conditions

Authors

Cabebe, G. - Presenter, Worcester Polytechnic Institute
Voltmer, S., Worcester Polytechnic Institute
Farny, N., Worcester Polytechnic Institute

Pseudomonas putida is a Gram-negative rhizobial bacterium capable of soil bioremediation and industrial biosynthesis. *P. putida* has a very diverse metabolism allowing it to perform these roles, but much is still unknown about its transcriptional regulation. This research aims to address the transcriptional programs that contribute to *P. putida*'s resilience and to develop a synthetic genetic construct to increase the survival rate of genetically engineered *P. putida* in the complex soil environment. To identify genes with consistent stable transcription, we performed a meta-analysis on *P. putida* KT2440 transcriptomic datasets and determined genes with low log-fold changes. A rank-based analysis was then carried out to determine highly expressed genes. The resulting top gene was found downstream of the *lapA* promoter region, the largest gene in *P. putida* whose transcriptional regulation is unclear and promoter region highly complex. Another key question is whether *P. putida* change their TSS selection or operon usage in response to the environment or competition from the native microbiome. We are currently investigating *P. putida*'s gene regulation by comparing operon-based expression in LB versus soil-extracted solubilized organic matter (SESOM), and will elucidate the differences between short Illumina and long Nanopore RNA-seq in the context of operons prediction. Development of these methods will enable us to dissect the role of the soil metabolic environment from the effects of competition from the microbial community in the context of genetic regulation. Our studies aim to uncover more about the relationship between environmental changes and the regulatory gene network of *P. putida*.

Targeted Reduction of Antibiotic Resistance Via Mobilizable Genetic Payloads in Klebsiella Pneumoniae

Authors

Schwanz, L. - Presenter
Lee, J., Columbia University
Shneider, D., Columbia University
Wang, H., Columbia University
Brockmann, L., Columbia University
Antibiotic resistance is a growing cause of mortality worldwide due to the emergence and proliferation of resistance genes in bacterial pathogens. One such pathogen is *Klebsiella pneumoniae*, a notorious carrier of many resistance genes harbored on numerous extrachromosomal elements. Most concerning is the increase in carbapenem-resistant strains, which has drastically reduced effective treatment options. Current treatments escalate antibiotic use but fail to address resistance genes themselves. To address this problem, we developed a plasmid-based therapy to enter *K. pneumoniae* cells and eliminate carbapenemase resistance genes directly. This therapy combines TP114, a self-transmissible plasmid with CRISPR-Associated Transposases (CASTs) for targeted DNA insertion into custom genomic loci, and I-C/Cas3 for targeted DNA cutting and bidirectional digestion. These tools together constitute a system that can effectively target and degrade extrachromosomal elements that harbor antibiotic resistance genes. Further, targeted insertion of I-C/Cas3 by CAST eliminates genomic copies of antibiotic resistance genes and prevents re-entry of resistance elements, thus preventing future acquisition of resistance. While eliminating target genes, this approach seeks to avoid killing all recipient cells of the broad-host TP114 plasmid, aiming to reduce further dysbiosis caused by off-target cutting events. This novel therapy class offers the unprecedented chance to selectively reduce antibiotic resistance of constituents within a bacterial community by physical removal of putative resistance genes. Future work will test various combinations of these elements to achieve the optimal resistance clearance payload.

Building a Synthetic Microaerobic Co-Culture for Upgrading C1 Substrates Using a High Throughput Bioreactor Platform.

Authors

Gasparrini, W. - Presenter, University of Connecticut
Zhou, G., Northeastern University
Jatkar, D., Northeastern University
Stohr, A., University of Delaware
Woolston, B., Northeastern University
Many important microbial consortia, such as the Human gut microbiota, consist of both aerobes and anaerobes that experience fluctuating levels of oxygen, yet little is known about how oxygen availability impacts community dynamics and metabolic interactions. We have developed a synthetic aerobe-anaerobe co-culture of E. coli and Clostridium ljungdahlii, in order to study these novel interactions. In this system, E. coli maintains the environment at steady low-oxygen conditions that support growth of C. ljungdahlii, while respiring on the fermentation byproducts of C. ljungdahlii. We have successfully established and analyzed the co-culture heterotrophically in both batch reactors and chemostats, and recently have shown that C. ljungdahlii can grow autotrophically under these conditions, despite using the highly oxygen-sensitive Wood-Ljungdahl pathway for CO_2 fixation. To scale these experiments, we have developed low-cost parallel mini-bioreactor platform in which gas mixes with different oxygen concentrations can be sparged to individual vials, and are now using this to understand the impact of varying environmental conditions on each microbe. Beyond understanding community dynamics, computational modeling suggests microaerobic co-culture of a mixed aerobic/anaerobic community could be broadly useful in metabolic engineering, enabling the coupling of high yield associated with anaerobic metabolism with the productivity of aerobic metabolism in a single reactor through cross-feeding. We present recent progress toward engineering our co-culture for continuous carbon-negative production of high-value chemicals from CO_2/H_2 to exemplify this concept. This work provides fundamental insight into the role of oxygen in microbial consortia, as well as a platform for analyzing impact of atmospheric composition.

Engineering the Gut Microbiome to Degrade Meat-Derived Sialic Acids

Authors

Phandanouvong-Lozano, V. - Presenter, University of Pittsburgh
Wolf, A., University of California Berkeley

Sialic acids serve as nutrient sources for commensal anaerobes in the gut microbiome, yet the metabolic pathways involved are poorly characterized. Sialic acids can be derived from meat food sources or host intestinal cells; these sugars are found at terminal positions of cell membrane glycans playing crucial roles in cell-to-cell communication and immune signaling. The meat-derived sialic acid Neu5Gc (n-glycolylneuraminic acid) has been linked to adverse immune responses associated with inflammatory disorders, atherosclerosis, and cancer. Our research seeks to identify or engineer native gut bacteria as putative probiotics capable of degrading Neu5Gc; and thus, mitigate related diseases. Humans cannot convert Neu5Ac (n-acetylneuraminic acid) into Neu5Gc due to an inactivated *cmah* hydroxylase gene, therefore Neu5Gc is only introduced through meat consumption. Although bacterial sialidases that release sialic acids residues from glycan chains can prevent Neu5Gc-incorporation into the human mucosa, free-sialic acids can be used by opportunistic pathogens triggering intestinal dysbiosis. Publicly annotated genomes from commensals such as *Hungatella* and *Bacteroides* encode putative sialic acid catabolic genes similar to the *nanAKE* pathway reported in *E. coli*; however, their specificity for Neu5Ac versus Neu5Gc consumption remains unknown. Through extensive screening of existing isolates, we have identified bacterial taxa with different specificities for Neu5Ac and Neu5Gc consumption. Following whole genome sequencing, selected bacterial taxa carrying putative sialic acid-catabolic genes are being characterized in order to identify enzymatic specificity to Neu5Ac or Neu5Gc. Our findings pave the way for the formulation of candidate probiotics that might contribute to mitigate health risks associated with meat consumption.

Unraveling Quantitative Relationships between Intracellular Phase Separation and Gene Expression through Single-Cell Analysis

Authors

Lee, D. - Presenter, Duke University
You, L., Department of Molecular Genetics and Microbiology
Dai, Y., Duke University
Chilkoti, A.

After the first observation of spontaneous occurrence of phase separation (PS) a living cell, the ubiquity of PS in a living system has been demonstrated. While our biophysical understanding of PS has drastically improved, its biological functions are much less known. It has been suggested that PS may regulate gene expression by selectively sequestering biomolecules into microcompartments formed by PS, and their concentrations inside the microcompartments are greatly elevated. Hence, if sequestered biomolecules are transcription factors or other molecules involving in gene expression, the increased concentrations are expected to enhance their downstream gene expression. However, past studies reported contradictory results. In order to elucidate quantitative relationships between PS and gene expression, we employed a single-cell fluorescence imaging methodology to establish quantitative correlations between phase separation and gene expression in E. coli. Specifically, we expressed a wide range of transcription factors (TFs) fused with a fluorescent protein (FP) and a PS-inducing motif (PSM) so that TFs undergo PS. By observing expression levels of another FP induced by a TF at the single-cell level, we can construct a dose-response curve between TF and FP in a high-throughput manner. By analyzing how PS occurrence affects a dose-response curve, we can quantitatively establish their correlations. By constructing dose-response curves with different PSMs with distinctive PS behaviors, the proposed screening method will provide a high-throughput way to characterize a variety of PSMs so that we can employ them as a new genetic part to control gene dynamics in a living cell.

AUTHOR INDEX